Environmental Science, Engineering and Technology

Environmental Science, Engineering and Technology

Flow Diagrams Applied to Microalgae-Based Processes
Ihana Aguiar Severo, PhD (Author)
2022. ISBN: 978-1-68507-742-6 (eBook)

Ecological Footprints: Management, Reduction and Environmental Impacts
Armano den Hartogh (Editor)
2022. ISBN: 978-1-68507-548-4 (Softcover)
2022. ISBN: 978-1-68507-606-1 (eBook)

Understanding Abiotic Stresses
Vishnu D Rajput, Krishan K. Verma and Tatiana M. Minkina (Editors)
2022. ISBN: 978-1-68507-508-8 (Hardcover)
2022. ISBN: 978-1-68507-552-1 (eBook)

An Innovative Approach of Advanced Oxidation Processes in Wastewater Treatment
Maulin P. Shah, PhD (Editor)
2021. ISBN: 978-1-68507-235-3 (Hardcover)
2021. ISBN: 978-1-68507-343-5 (eBook)

Environmental Management: Ecosystems, Competitiveness and Waste Management
Miguel Fischer (Editor)
2021. ISBN: 978-1-68507-019-9 (Softcover)
2021. ISBN: 978-1-68507-059-5 (eBook)

More information about this series can be found at
https://novapublishers.com/shop/environmental-management-ecosystems-competitiveness-and-waste-management/

Fisnik Osmani and Atanas Kochov

Applications of AHP Methodology for Decision-Making in Cleaner Production Processes

www.novapublishers.com

Copyright © 2022 by Nova Science Publishers, Inc.
https://doi.org/10.52305/CEQH2502

All rights reserved. No part of this book may be reproduced, stored in a retrieval system or transmitted in any form or by any means: electronic, electrostatic, magnetic, tape, mechanical photocopying, recording or otherwise without the written permission of the Publisher.

We have partnered with Copyright Clearance Center to make it easy for you to obtain permissions to reuse content from this publication. Simply navigate to this publication's page on Nova's website and locate the "Get Permission" button below the title description. This button is linked directly to the title's permission page on copyright.com. Alternatively, you can visit copyright.com and search by title, ISBN, or ISSN.

For further questions about using the service on copyright.com, please contact:
Copyright Clearance Center
Phone: +1-(978) 750-8400   Fax: +1-(978) 750-4470   E-mail: info@copyright.com.

NOTICE TO THE READER

The Publisher has taken reasonable care in the preparation of this book, but makes no expressed or implied warranty of any kind and assumes no responsibility for any errors or omissions. No liability is assumed for incidental or consequential damages in connection with or arising out of information contained in this book. The Publisher shall not be liable for any special, consequential, or exemplary damages resulting, in whole or in part, from the readers' use of, or reliance upon, this material. Any parts of this book based on government reports are so indicated and copyright is claimed for those parts to the extent applicable to compilations of such works.

Independent verification should be sought for any data, advice or recommendations contained in this book. In addition, no responsibility is assumed by the Publisher for any injury and/or damage to persons or property arising from any methods, products, instructions, ideas or otherwise contained in this publication.

This publication is designed to provide accurate and authoritative information with regard to the subject matter covered herein. It is sold with the clear understanding that the Publisher is not engaged in rendering legal or any other professional services. If legal or any other expert assistance is required, the services of a competent person should be sought. FROM A DECLARATION OF PARTICIPANTS JOINTLY ADOPTED BY A COMMITTEE OF THE AMERICAN BAR ASSOCIATION AND A COMMITTEE OF PUBLISHERS.

Additional color graphics may be available in the e-book version of this book.

Library of Congress Cataloging-in-Publication Data

ISBN: 978-1-68507-882-9

Published by Nova Science Publishers, Inc. † New York

Contents

Preface vii

Acknowledgments ix

Chapter 1   **Introduction** 1
                     1.1. Concept of Sustainable Development 3
                     1.2. Indicators of Sustainable Development –
                     Environmental Dimension 5
                     1.3. Indicators of Sustainable Development –
                     Technical Dimension 7
                     1.4. Indicators for Sustainable Development –
                     Economic Dimension 8
                     1.5. Indicators of Sustainable Development –
                     Social Dimension 9

Chapter 2   **Analytic Hierarchy Process and Multi-Criteria
                     Decision-Making** 11
                     2.1. Mathematical Definition of MCDM 11
                     2.2. Review of Eleven MCDM Methods Identified
                     throughout the Analyses 14
                     2.3. Differences and Similarities Between Methods
                     of Multi-Criteria Decision-Making 15
                     2.4. Phases in the Process of AHP Application 18
                     2.5. Multi-Objective (Multi-Goal) Programming 21
                     2.6. Creating the Hierarchy of the Problem 22
                     2.7. Hierarchy of the Problem Along with
                     the Alternatives 28

Chapter 3   **Concept and Application of Resource Efficient
                     and Cleaner Production (RECP)** 31
                     3.1. The Concept and Indicators by UNIDO 34

Chapter 4	**Defining Indicators for Decision and Policy Making for Contributing to Sustainable Development**	...37
	4.1. Making Decisions	...37
	4.2. Pyramid of Information, Simple and Complex Indicators	...39
	4.3. Expert Choice Software	...40
Chapter 5	**Applied Methodology: Process and Method of Multi-Criteria Decision-Making (MCDM), Analytic Hierarchy Process (AHP) and Multi-Objective Programming (MP)**	...49
	5.1. Calculation of the Weight Factors of the Indicators	...49
	5.2. Multi-Objective Programming Approach Analysis	...74

Discussion of Analysis and Conclusion ...83

The Recommended Application of the AHP Method ...87

References ...91

About the Authors ...97

Index ...99

Preface

We would like to thank all the readers for their patience in reading this book. The decision-making process in cleaner production is very important, taking into account the complexity and multi-criteria of decision-making in this area, when applying the AHP method.

This book addresses the theoretical and scientific foundations and practical research of technologies in clean production processes – based on multi-criteria decision-making. The analyzed/selected model is a realistic and appropriate model for the field of research. Comparisons are made with traditional decision-making methods and new advanced scientific methods are incorporated. The book will be of interest to engineers and researchers who focus on multi-criteria decision-making and research in clean production processes, as well as students and professors at universities that deal with this field.

Fisnik Osmani, PhD
Assistant Professor,
Faculty of Mechanical Engineering and Computers,
University "Isa Boletini",
Mitrovica, Republic of Kosovo

Atanas Kochov, PhD
Professor,
Mechanical Engineering,
Ss. Cyril and Methodius University,
Skopje, North Macedonia

Acknowledgments

We would like to acknowledge with gratitude our families, colleagues and host universities for all their support in our research work and professional development. They kept us going with their support, help, patience and encouragement to work, explore, research, and in our writing.

Chapter 1

Introduction

The provision of sufficient energy and its rational exploitation make up one of the biggest challenges in the global sphere, especially in developing and industrializing countries. The Western Balkan countries face a situation that is not favorable in the logic of sustainable development. However, it must be acknowledged that positive developments have taken place in recent years, taking into account that we are talking about an area that for years has been an area of open conflict.

Last year, there were important changes in the energy supply sector, which have important implications for energy planning. The Western Balkan region faces critical energy and development choices that will impact the energy supply available to meet its basic needs and provide economic growth. These choices will also impact the health of the population, determine the job creation potential of the energy sector, and impact the wider regional role that the Western Balkan Countries may play in the European Union.

Kosovo, as a country in transition with an "industry in the making", needs to seriously consider the above-mentioned issue. The industry should be subject to professional analysis (the application of resource efficiency and cleaner production) in order to achieve the proper use of resources. Consequently, as a result of increased resource efficiency, greater effectiveness will be achieved by enterprises. This will undoubtedly enable the achievement of cleaner production – and the launch of the RECP application in Kosovo – as the only country in the region that has not made any move in this regard.

Key and important sectors in general in the Western Balkan Countries, including Kosovo, are Energy, Mining, ICT, Health, Waste Management and Recycling, Food, Beverages, Tobacco/Textiles, Leather, Shoes/Wood Processing/Paper, Graphics, Publishing/Plastics & Tires/Metallurgy & Metalworking/Building Materials - factors which will affect the industrial development of this area. Undoubtedly, these industries have already changed their profile over the years.

The industry players, irrespective of all the above sectors, are large consumers of energy, environment polluters, and waste producers on different scales. Given the current situation, we have analyzed the possibility of applying Resource Efficient and Cleaner Production (RECP) in Kosovo. Through RECP, we have managed to see in which direction to move, what are the steps needed to achieve a level within the allowed norms of rational energy consumption, to minimize environmental pollution, minimize waste, and maximize the profits of enterprises, which can lead to the creation of more jobs.

Through this analysis, we have also established concrete steps that should be undertaken to apply RECP, and thus come closer to the criteria applied by the most developed countries of the European Union.

Based on the aspiration of the countries of the region, their individual functional form, and the challenges that they face, there is no doubt that the common factor of achieving their objectives is sustainable development based on resource efficiency and cleaner production.

The analysis which we have done promotes a stable development based on Resource Efficient and Cleaner Production (RECP) in the countries of the region and especially Kosovo – in cleaner production and in the energy sector. This analysis based on RECP has been tested with the method of Multi-Criteria Decision-Making – Analytic Hierarchy Process for all the levels of decision-making. It is based on the four main pillars of sustainable development, respectively environment, technical, economic and social.

The identification of indicators for the four pillars is of utmost importance to achieve a real reflection through which it is possible to measure progress and development based on the main components of a stable development based on RECP.

The entire analysis has been done with special care in accordance with the main results of the United Nations Conference for the Environment and Development which took place in 1992 [1]. From there it is derived that stable development and the indicators which derive from it will serve as assistance for countries and competent decision-making institutions.

In order to fulfill with ease our objective of defining the indicators for stable development for the Western Balkan countries, we have compiled indicator lists for the whole region. This listing has been done by collecting accurate and reliable data from state statistical agencies.

The classification of stable indicators is listed for each country individually, by delving into the main themes and sub-themes, which are treated based on the indicators that have been identified by the United Nations (the 17 goals of the UN) [2], and by adapting also the data from the respective countries according to the analysis.

The orientation on which indicators to consider in our analysis has been carried out by combining indicators from the UN and by adding other indicators also, taking into consideration the wide range of indicators that are part of the UN treatment.

1.1. Concept of Sustainable Development

Sustainable development will be addressed at the Western Balkans level in order to address the resource efficiency and enabling conditions for cleaner production, focusing on the Republic of Kosovo and aiming to establish a "cleaner" and more stable energy sector. The term "sustainable development" has a number of definitions and interpretations, but in this work the meaning of the term is in accordance with the United Nations (UN) concept.

"Sustainable development is development that meets the needs of the present, without compromising the ability of future generations to meet their own needs" [3]. It contains within it two key concepts: the concept of needs, in particular the essential needs of the world's poor, to which overriding priority should be given; and the idea of limitations imposed by the state of technology and social organization on the environment's ability to meet present and future needs [4], Development: the act or process of developing; growth; progress.

The term sustainable development began to gain wide acceptance in the late 1980s, after its appearance in "Our Common Future", also known as the Brundtland Report [5]. The result of a UN-convened commission created to propose "a global agenda for change" in the concept and practices of development, the Brundtland Report signaled the urgency of re-thinking the ways of living and governing.

To "responsibly meet humanity's goals and aspirations" will require new ways of considering old problems as well as international co-operation and co-ordination. The World Commission on Environment and Development, as it was formally called, sought to draw the world's attention to "the accelerating deterioration of the human environment and natural resources, and the consequences of that deterioration for economic and social development" [6, 7]. In establishing the commission, the UN General Assembly explicitly called attention to two important ideas: the well-being of the environment, of economies and of people is inextricably linked.

Lately, with the Resolution adopted by the General Assembly on 25 September 2015 "Transforming our world: the 2030 Agenda for Sustainable

Development", the countries of the entire world have reached a decision through the approval of a list with 17 goals [8, 9].

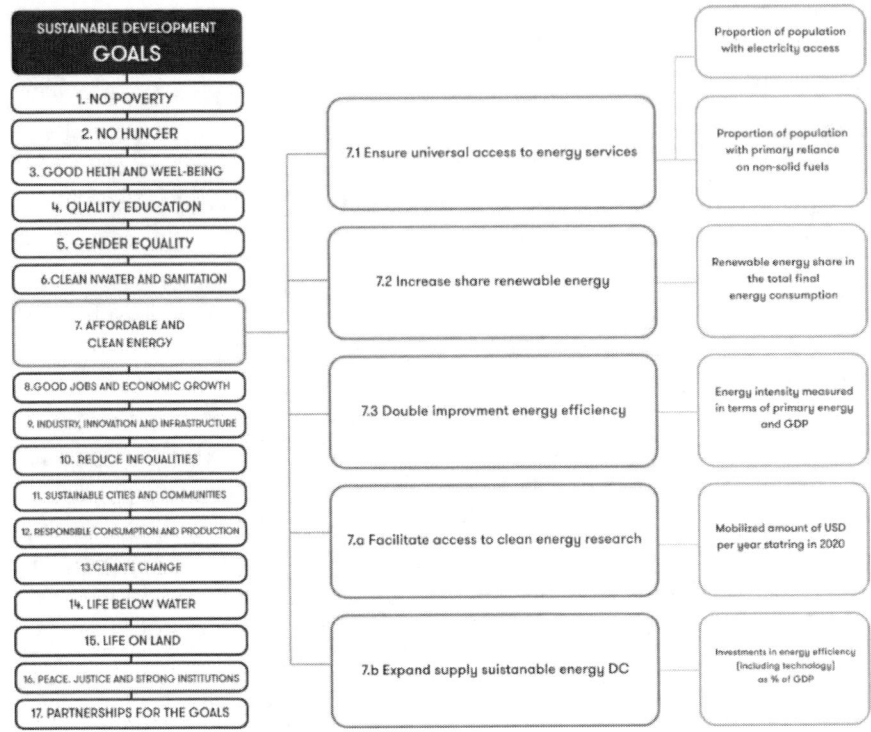

Figure 1. Sustainable Development Goals: Goal 7, Affordable and clean energy (Ensure access to affordable, reliable, sustainable and modern energy for all).

The commission was constituted by the UN in 1984, under the chairmanship of Gro Harlem Brundtland, the former Prime Minister of Norway, with a mission to address growing concern over the "accelerating deterioration of the human environment and natural resources and the consequences of that deterioration for economic and social development." The Brundtland Commission was officially dissolved in December 1987 after releasing "Our Common Future," also known as the Brundtland Report, in October 1987, a document which coined and defined the meaning of the term "Sustainable Development". The Brundtland Report laid the groundwork for the convening of the Earth Summit in Rio de Janeiro five years later. The Report strongly influenced the subsequent initiatives towards sustainable development across the world [10, 11, 12]. Of the 17 goals presented, in Figure

1 there is a special focus on goal No. 7, concerning the provision of affordable and clean energy for all.

Also, the United Nations General Assembly ad opted the Resolution 70/1, Transforming our World: the 2030 Agenda for Sustainable Development on 25th September 2015. This document lays out the 17 Sustainable Development Goals, which aim to end poverty and hunger, protect human rights and human dignity, to protect the planet from degradation, and foster peace [13].

The concept of sustainable development can be interpreted in many different ways, but at its core is an approach to development that looks to balance different, and often competing, needs against an awareness of the environmental, social and economic limitations with which society is faced.

1.2. Indicators of Sustainable Development – Environmental Dimension

The environmental dimension is a pillar which is essential for sustainable development. In this respect, the main role is played by the environmental protection and indicators which contribute in this direction. Today, almost all the analyses, research and the majority of the literature [14, 15, 16] unanimously conclude that humanity is sliding to a level of environmental degradation which is seriously endangering the planet Earth.

The pollution, which has passed all limits, the global warming and the increase in temperatures, climate change, uncontrolled increase of deforestation, high pollution from heavy industry, pollution from electrical energy production facilities (which is most common in the Western Balkan countries) are just some of the indicators that confirm the damage to the environment.

In order to understand the environmental trends in the Western Balkan countries and identify key indicators, various indicators are first selected, including: atmosphere, fresh water, and pollution from heavy industry, with a special emphasis on electrical energy production (TPP Kosovo B).

The definition of indicators for sustainable development in the environmental dimension has provided a clear and real picture, including many parties. Also, the measures that should be taken by each country and for all the Western Balkan countries as a whole should be in a direction in which all the activities that degrade the environment will be halted, or at least their intensity decreased. From the indicators, it is expected to give a clear picture of the direction that should be taken for the purposes of reasoned and objective decision-making for stable development. Defining the indicators for sustainable development in respect to the dimension of the environment is based on the strategic national documents in this area concerning the region, the projects implemented by relevant institutions, etc. Having in mind these issues, the following sources of statistical indicators are used in considering this group of indicators:

- Strategic policy documents for protection of the environment
- Number and type of pollutants
- Number of reforested hectares
- Number of recycling facilities established
- Number of public awareness campaigns and targeted population
- Number of metric tons of garbage removed from rivers, lakes and forests

Based on the analysis of relevant documents, strategies, reports, scientific papers and other sources, in this book the following indicators that influence the environment [17, 18, 19] are identified:

- Resource efficiency
- CO_2 emission
- Waste treatment
- PM emission
- Contamination of soil
- Landscape changes
- Energy efficiency
- Efficiency of materials
- CO_2 - kWh
- Number of landfills and number of wastewater treatment plants
- PM 10 – PM 2.5
- Number of contaminants
- m^2 area/MW installed capacity for different technologies

The indicators on the environmental dimension are expected to be crucial sources of information in compiling Environmental Impact Assessment reports, which are nowadays very common and necessary tools for improving the overall impact on the environment of undertakings that originate from the energy sector and the industry in general.

1.3. Indicators of Sustainable Development – Technical Dimension

In the United Nations Sustainable Development Goals, the seventh objective has to do with Affordable and Clean Energy [20]. In this component, the focus is on defining the technical factors that relate precisely to the provision of clean energy, and the possibility of providing the supply. Also, the potentials of the countries of the region, in this case Kosovo, Albania and Macedonia, in the supply of clean energy are herein analyzed in terms of combined heat and power (CHP), and the use of natural gas, biomass, wind power, hydro power, and geothermal energy. In the domain of technical factors other indicators are identified that are addressed in this analysis, which are further listed in the hierarchy of the problem [21, 22, 23, 24]:

- Securing clean energy
- Security of supply
- Availability of energy
- Availability of know-how
- Combined heat and power plants
- The use of natural gas
- The use of biomass
- The use of wind power
- The use of hydro power
- Geothermal energy use
- Import dependency of energy commodities
- Geo-political issues
- Natural disasters
- Primary domestic energy reserves
- Stochastic nature
- Number of foreign direct investments
- Number of educated engineers
- Number of foreign companies
- Number of educated skilled technicians

1.4. Indicators for Sustainable Development – Economic Dimension

It is natural and logical that the economic dimension is the segment that has the greatest impact on sustainable development. Consequently, it is the dimension that is the most extensively treated in maximizing the benefits and using resources which have effects on sustainable development policy and decision-making [25].

In this analysis, the indicators for sustainable development listed under the economic dimension have special importance as a main pillar for drawing up the complete platform for sustainable development. The whole analysis is divided into sub-topics through which the economic sustainability of the Western Balkan countries as a whole is evaluated, with special emphasis on Kosovo. Apart from indicators such as production, the level of employment and the unemployment rate, the main emphasis was on the definition of indicators which are related to sustainable development in the energy sector (sources of electrical and thermal energy, as well as alternative sources) and which can provide a sustainable supply. The goal of such an approach is to improve the management and development issues concerning the energy sector. Indeed, it is well known that the main challenge for the economic development of each of the Western Balkan countries is the energy sector. It is the most important sector for economic development as one of the preconditions set by the EU before proceeding with the next steps [26]. Therefore, in this work it is decided to treat the energy sector as one of the most important concerning the economic dimension.

Accordingly, the indicators that are analyzed and the results that are obtained will serve to give a real picture, precise and independent of the level of progress that the Western Balkan countries have achieved in their economic development, especially in the field of energy, its production, distribution, and consumption, including strategic actions to use alternative sources. Hence, in the analysis of this group of indicators, the following categories of information were selected as a base [27, 28]:

- Number of new power plants
- The categories of new power plant (wind, hydro, coal, other renewable) investments
- The amount of investments
- Percentage of energy share in GDP

- Nominal share of energy in GDP

The indicators on the economic dimension will enable the presentation of the structure of investments in the energy sector as well as their overall impact on the economic development of the region and in each individual country.

The economic indicators that are analyzed in this case are [29, 30]:

- IRR – Internal Rate of Return
- Reducing energy poverty (% of household income spent on energy bills)
- Economic growth
- Investment cost (Euro/MW)
- Operational cost
- Environmental cost (externalities)

1.5. Indicators of Sustainable Development – Social Dimension

There is ever more discussion/analysis and treatment of the indicators for sustainable development, focusing on the pillars of the economic and environmental indicators. Despite that, awareness of the social dimension should always be present, and attention to these problems is still required.

In this context, vital topics that are related with the social aspect are analyzed in this book, taking into consideration that the Western Balkan countries as a whole face similar, if not identical, issues [31, 32]. The aspects of education, residence, healthcare, and security are identified, along with one of the largest issues that is heavily burdening the integration of the Balkan countries – migration.

The social dimension is regarded as one of the main factors for sustainable development, since it has a special importance and influence over the environmental and economic dimensions. To put it simply, the social dimension is directly linked to sustainability and the quality of life, which is the main focus of the environmental and economic dimensions. With respect to this finding, the following aspects of the social dimension are considered:

- Number of new jobs created
- Number of energy-efficient houses (new and renovated)
- Indicators on energy housing efficiency

- Number of new enrolments in educational programs (vocational and post-graduate)
- Number of illnesses reduced

As concerns the above listed indicators, it is expected to show improvements as a result of the new sustainable investments and these will reflect changes in society in terms of the social dimension [33, 34, 35].

The social indicators analyzed in this work are:

- Safety and health
- Good governance
- Quality of life
- Air quality (Average level of PM)
- Number of deaths due to air pollution
- Voice and accountability (VA)
- Political stability and absence of violence (PV)
- Government effectiveness (GE) and regulatory quality (RQ)
- Rule of law (RL)
- Control of corruption (CC)
- GDP per capita
- HDI - human development index

Chapter 2

Analytic Hierarchy Process and Multi-Criteria Decision-Making

Multi-Criteria Decision Analysis has seen an incredible amount of use over the last several decades. Its role in different application areas has increased significantly, especially as new methods develop and as old methods improve.

This work analyzes several common methods of MCDM and determines their applicability in different situations by evaluating the disadvantages and advantages. Firstly, a comprehensive review of the literature is performed in order to prepare a summary of the most used methods as well as to conduct more research in terms of their features.

In addition to applying single MCDM methods to real-world decisions, the progression of technology over the past couple of decades has allowed for more complex decision analysis methods to be developed. This experimentation with combined multi-criteria decision-making methods has provided a whole new approach to decision analysis.

2.1. Mathematical Definition of MCDM

Multi-Criteria Decision-Making (MCDM) can be defined as a methodology that facilitates the evaluation of real-world situations, based on various qualitative and/or quantitative criteria in certain, uncertain, or risky environments to suggest a suitable course of action, choice, strategy, or policy among the available options. This problem becomes even more complex when conflicting and non-commensurable criteria are present and when a significant number of decision-makers are involved [36].

The main components of the MCDM are:

- the attributes, represented through the corresponding attributes set X,
- the criteria or objectives and their relevant indicators – represented through the corresponding criteria set S, and

- the alternatives – represented through the corresponding alternatives set A

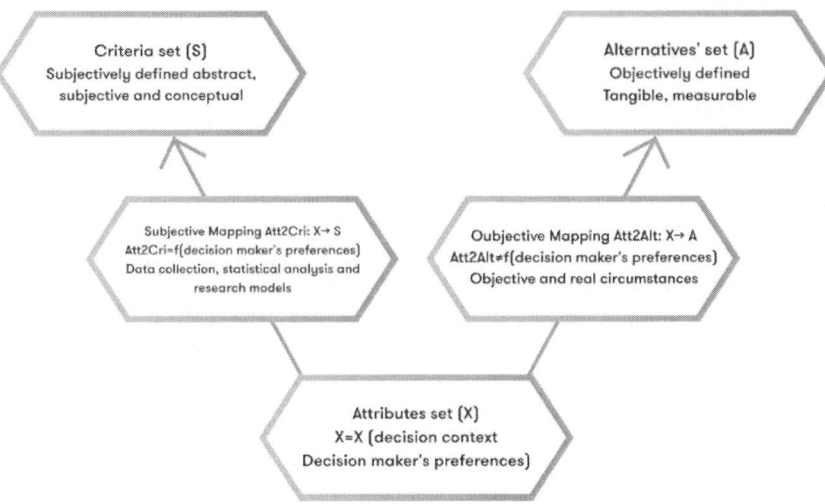

Figure 2. Components of MCDM and their interrelations.

The methodology is convenient for breaking down a complex, unstructured situation into its component parts, then arranging these parts into a hierarchical order (criteria, sub-criteria, indicators) and assigning numerical values from 1 to 9 to subjective judgments on the relative importance of each criterion/indicator using pair-wise comparison. Saaty suggests that hierarchies are to be limited to six levels, with nine items per level. This is based on the psychological finding that people can consider 7+/-2 items simultaneously (Miller, 1956) [36, 37, 38].

Solving/addressing the mathematical problem of MCDM may be required through this formulation. Further details are thoroughly elaborated in Saaty (1986, 1990) (1986) [39, 40, 41]. The decision (or the goal achievement) matrix, *MxN* **X**, aggregates the complete problem-related information and is a basis for the problem solution. In the thus defined decision matrix, we consider that the subjective mapping of the attributes set (*X*) onto the criteria set (*S*) has already been performed, i.e., *N* is the number of the mapped criteria relevant for the calculation of weights and thus the decision-making.

$$X = \left| x_{ij} = f_j(A_i) \right|_{MxN}, \quad i = 1, M, \quad j = 1, N \quad (1)$$

$$\begin{bmatrix} & X_1 & X_2 & \dots & X_j & \dots & X_N \\ & w_1 & w_2 & \dots & w_j & \dots & w_N \\ A_1 & x_{11} = f_1(A_1) & x_{12} = f_2(A_1) & \dots & x_{1j} = f_j(A_1) & \dots & x_{1N} = f_N(A_1) \\ \dots & \dots & \dots & \dots & \dots & \dots & \dots \\ A_i & x_{i1} = f_1(A_i) & x_{i2} = f_2(A_i) & \dots & x_{ij} = f_j(A_i) & \dots & x_{iN} = f_N(A_i) \\ \dots & \dots & \dots & \dots & \dots & \dots & \dots \\ A_M & x_{M1} = f_1(A_M) & x_{M2} = f_2(A_M) & \dots & x_{Mj} = f_j(A_M) & \dots & x_{MN} = f_N(A_M) \end{bmatrix}$$
(2)

where M and N are the number of alternatives and criteria, respectively, while $x_{ij} = f_j(A_i)$ indicate the value of the j x criterion with respect to the alternative A_i.

$$S = \{f_1, f_2, \dots, f_N\} \tag{3}$$

is the set of criteria defined as

$$(\forall x \in X)(\exists f(x) \in S): X \mapsto S = \{f(x) | x \in X\} \tag{4}$$

where

$$X = \{x | g(x) \le 0\} \text{ and } g(x) \le 0 \tag{5}$$

is the problem-related set of attiributes and the corresponding vector of constraints, respectively.

$$A = \{A_1, A_2, \dots, A_M\} \tag{6}$$

is the set of the feasible alternatives identified. A weighting factor, w_j, can be associated with each criterion, indicating its importance. Then, the "best" solution to a MCDM problem can be defined as

$$\max/_x \min U(f) = \sum_{i=n}^{N} w_i \cdot u_i(f_j(x)) \tag{7}$$

where $U_i(f)$ is the overall utility function calculated for the alternative A_i while w_j and u_j are the weighting factor and the utility related to a particular criterion and the corresponding alternative A_i, [36, 42].

2.2. Review of Eleven MCDM Methods Identified throughout the Analyses

As concerns the MCDM methods, it can be concluded without doubt that there is no universally applicable method that will deliver the optimal results for any problem or application. Figure 3 presents the eleven most common MCDM methods in the literature, as well those most used, in particular when it comes to solving problems related with the energy and transportation infrastructure, and with the adoption of certain public policies, strategic documents, etc.

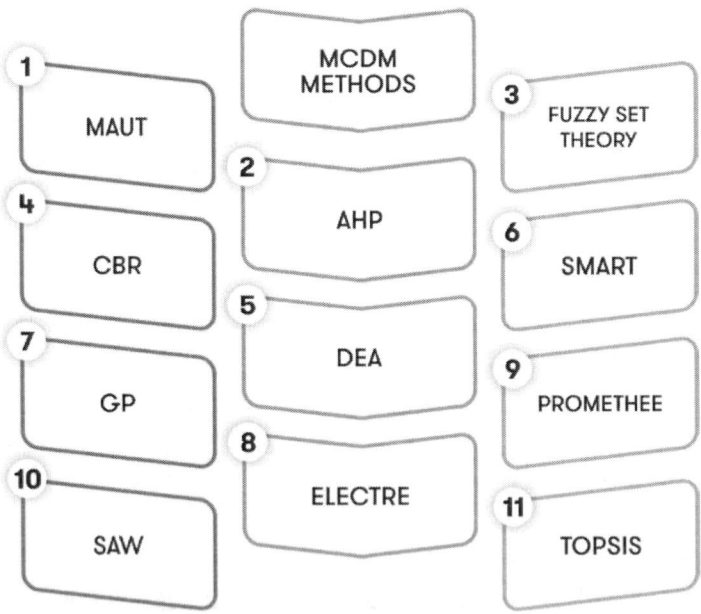

Figure 3. MCDM methods.

The abbreviations in Figure 3 indicate the following: Multi-Attribute Utility Theory, Analytic Hierarchy Process, Fuzzy Set Theory, Case-based

Reasoning, Data Envelopment Analysis, Simple Multi-Attribute Rating Technique, Goal Programming, SAW, PROMETHEE, ELECTRE, [43, 44].

The advantages and disadvantages of the eleven methods described in the literature [40, 41] are discussed below. Their common application will also be analyzed in order to find out whether a certain interrelation can be established in terms of the method's applicability and these advantages and disadvantages. The conclusion confirms that certain methods are better for certain situations, and vice versa.

AHP is "a theory of measurement through pairwise comparisons and relies on the judgments of experts to derive priority scales" [40, 41].

The AHP was developed first by Saaty (1980) as a method for solving complicated and unstructured problems that may have interactions and correlations among different objectives and goals. It is one of the more popular methods of MCDM and has many advantages, as well as disadvantages. One of its advantages is its ease of use. It has a hierarchy structure which is easily adaptable and adjustable in accordance with the analyzed problem [40, 41].

The need to make the right decisions requires the application of accurate, true and tested methods that have proven to provide reliable results. In our analysis we applied the Multi-Criteria Decision-Making process, the AHP hierarchical analytical method. In the field of decision-making, MCDM has seen widespread and effective application, and has consequently proved to be a method that has provided accurate, reliable results supported by mathematical verification and analysis.

2.3. Differences and Similarities Between Methods of Multi-Criteria Decision-Making

As already elaborated, there are a number of multi-criteria decision-making (MCDM) methods that are used in different applications. It should be pointed out that their modification and transformation are a continuous process, i.e., they have changed with the aim of adapting to different application areas.

Multi-criteria decision analysis has seen an incredible amount of use over the last several decades. Its role in different application areas has increased significantly, especially as new methods develop and as old methods improve [36].

Table 1 presents the advantages, disadvantages and application fields for all the methods that come under MCDM, as shown above in Figure 3.

Table 1. Summary of MCDM methods [36, 39]

Method	Advantages	Disadvantages	Areas of Application
Multi-Attribute Utility Theory (MAUT)	Takes uncertainty into account and can incorporate preferences	Needs a lot of input, preferences need to be precise	Economics, finance, actuarial, water, management, energy management, agriculture
Analytic Hierarchy Process (AHP)	Easy to use, scalable, hierarchy structure can easily be adjusted to fit many sized problems, not data-intensive	Problems due to interdependence between criteria and alternatives can lead to inconsistencies between judgment and ranking criteria, several ranks	Performance-type problems, resource management, corporate policy and strategy, public policy, political strategy and planning.
Case-Based Reasoning (CBR)	Not data-intensive, requires little maintenance, can improve over time, can adapt to changes in environment	Sensitive to inconsistent data, requires many cases	Business, vehicle insurance, medicine and engineering design
Data Envelopment Analysis (DEA)	Capable of handling multiple inputs and outputs, efficiency can be analyzed and quantified	Does not deal with imprecise data, assumes that all input and output are exactly known	Economics, medicine, utilities, road, safety, agriculture, retail and business problems.
Fuzzy Set Theory	Allows for imprecise input, takes into account insufficient information	Difficult to develop, can require numerous simulations before use	Engineering, economics, environmental, social, medical and management
Simple Multi-Attribute Rating Technique (SMART)	Simple, allows for any type of weight assignment technique, less effort by decision-makers	Procedure may not be convenient considering the framework	Environmental, construction, transportation and logistics, military, manufacturing and assembly problems
Goal Programming (GP)	Capable of handling large-scale problems, can produce infinite alternatives	Its ability to weight coefficients typically needs to be used in combination with other MCDM methods of weighting them	Production planning, scheduling health care, portfolio selection, distribution systems, energy planning, water reservoir management, scheduling, wildlife management

Method	Advantages	Disadvantages	Areas of Application
ELECTRE	Takes uncertainty and vagueness into account	Its process and outcome can be difficult to explain in layman's terms, outranking causes the strengths and weaknesses of the alternatives to not be directly identified	Energy, economics, environmental, water management and transportation problems
PROMETHEE	Easy to use, does not require assumption that criteria are proportionate	Does not provide a clear method by which to assign weights	Environmental, hydrology, water management business and finance, chemistry, logistics and transportation, manufacturing and assembly, energy, agriculture
Simple Additive Weighting (SAW)	Ability to compensate among criteria, intensive for decision-makers, calculation is simple, does not require complex computer programming	Estimates revealed do not always reflect the real situation, result obtained may not be logical	Water management, business and financial management
Technique for Order Preferences by Similarity to Ideal Solutions (TOPSIS)	Has a simple process, easy to use and program, the number of stops remains the same regardless of the number of attributes	Its use of Euclidean distance does not consider the correlation of attributes, difficult to weight and keep consistency of judgment	Supply chain management and logistics, engineering manufacturing systems, business and marketing, environmental, human resources and water resources management

2.4. Phases in the Process of AHP Application

By analyzing the process of the application of AHP, in solving a specific problem, the following four phases can be identified [45]:

- Structuring of the problem
- Data collection
- Evaluation of the relative weight
- Determination of the problem solutions

The first phase involves the disintegration of the decision-making problem into a series of hierarchical levels, each of which represents a smaller number of controllable attributes. AHP is based on mutual comparison of elements in a given hierarchical level relative to the elements of a higher level. As such, if we look closely at the general case of a hierarchy with three levels (goal – criteria – alternatives) (Figure 4), the criteria are compared relative to the goal, in order to determine their joint importance, and alternatives to each of the set criteria.

The data collection phase, on the other hand, is the second phase of the AHP method involving the collection and measurement of data. The procedure follows certain steps, including assigning a relative assessment in pairs with the attributes of a hierarchical level, for given attributes of the first and higher hierarchical level, then repeating the process for all levels of the hierarchy. To assign a weight, Saaty's nine-point scale is used, as shown in Table 2.

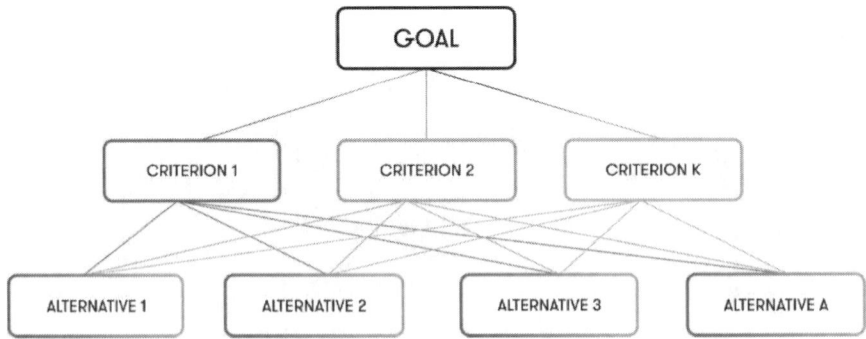

Figure 4. General hierarchy with three levels (goal – criteria – alternatives).

Table 2. The scale of relative priorities

Scale	Compare factor of i and j
1	Equally important
3	Weakly important
5	Strongly important
7	Very strongly important
9	Extremely important
2,4,6,8	Intermediate values between adjacent scales

The third phase of the AHP method is to estimate the relative weight. Based on the matrix A with elements a_{ij}, the priorities of the criteria, sub-criteria and alternatives are determined [46].

After determining the weights, their credibility should be established. This is done by determining the consistency of the matrix A. The matrix A, in the case of consistent (consequent) assessment with a_{ij} a_{ik} a_{kj}, satisfies the equation:

$$A_w = n \cdot w \tag{8}$$

or:

$$A = \begin{bmatrix} a_{11} & a_{12} & \cdots & a_{1n} \\ a_{21} & a_{22} & \cdots & a_{2n} \\ \cdots & \cdots & \cdots & \cdots \\ a_{n1} & a_{n2} & \cdots & a_{nn} \end{bmatrix} \quad A = \begin{bmatrix} \frac{w_1}{w_1} & \frac{w_1}{w_2} & \cdots & \frac{w_1}{w_n} \\ \frac{w_2}{w_1} & \frac{w_2}{w_2} & \cdots & \frac{w_2}{w_n} \\ \cdots & \cdots & \cdots & \cdots \\ \frac{w_n}{w_1} & \frac{w_n}{w_2} & \cdots & \frac{w_n}{w_n} \end{bmatrix} \cdot \begin{bmatrix} w_1 \\ w_2 \\ \cdots \\ w_3 \end{bmatrix} = n \cdot \begin{bmatrix} w_1 \\ w_2 \\ \cdots \\ w_3 \end{bmatrix}$$

$$\tag{9}$$

The characteristics of the matrix A:

$$a_{ij} = 1 \, ; a_{ij} = 1/a_{ij} \quad for \, i,j = 1,...,n \, ; \, \det A \neq 0 \tag{10}$$

The problem of determining the weights can be solved as a problem of solving a matrix equation, with the matrix columns w solution for eigenvalues λ different from 0, i.e.,

$$A \cdot w = \lambda \cdot w, \text{ or } \begin{bmatrix} a_{11} & a_{12} & \cdots & a_{1n} \\ a_{21} & a_{22} & \cdots & a_{2n} \\ \cdots & \cdots & \cdots & \cdots \\ a_{n1} & a_{n2} & \cdots & a_{nn} \end{bmatrix} \cdot \begin{bmatrix} w_1 \\ w_2 \\ \cdots \\ w_n \end{bmatrix} = \begin{bmatrix} \lambda_1 w_1 \\ \lambda_2 w_2 \\ \cdots \\ \lambda_n w_n \end{bmatrix} \qquad (11)$$

If the matrix A contains inconsistent assessments, the weight vector w can be obtained by solving the following equation:

$$(A - \lambda_{max1}) \cdot w = 0 \text{ if } \sum w_1 = 1 \qquad (12)$$

where λ_{max} is the largest eigenvalue of the matrix A:

$$\lambda_{max} = \frac{1}{n} \sum_{i=1}^{n} \frac{(Aw)_i}{w_i} \quad \lambda_{max} \geq n \qquad (13)$$

and the difference $\lambda_{max} - n$ is used in measuring the consistency of the assessment, or to calculate the index of consistency:

$$CI = (\lambda_{max} - n)/(n-1) \qquad (14)$$

Based on this index, we determine the index of inconsistency:

$$CR = CI / RI \qquad (15)$$

The variable RI in Equation (15) represents the so-called Random Index, and its values are chosen from Table 3.

Table 3. RI (Random Index)

n	1	2	3	4	5	6	7	8
RI	0	0	0.52	0.89	1.11	1.25	1.35	1.40

The value of $CR \leq 0.10$ indicates that the estimates for a and j are consistent. Otherwise, the evaluation should be repeated.

The last phase of the AHP method is finding the so-called composite normalized vector. Since the successive levels of the hierarchy are interconnected, a single composite vector of unique normalized weight vectors for the entire hierarchy is determined by multiplying the weight vectors of all successive levels. The composite vector is used to find the relative priority of the entities at the lowest (hierarchical) level, which allows the goals of the overall problem to be achieved.

2.5. Multi-Objective (Multi-Goal) Programming

Decision-making on many criteria, or MCDM, is implemented when there are contradictory criteria, that is to say, several criteria that matter but cannot be optimized at the same time. In order to exploit the advantages of the AHP application, the method of multi-objective [47, 48, 49] (multi-goal) linear programming is used. Multi-objective programming (MOP) or vectored optimization techniques address the simultaneous optimization of some of the objectives that are subject to a set of commonly linear constraints. When an optimization solution cannot be determined for some objectives, MOP is used to gain potential community solutions that are efficient solutions (Pareto Optimal) [50], rather than finding a single optimal solution. The elements of this efficient group are the possible solutions; there are no other possible solutions that can achieve the same or better performance for all objectives and in the best way for at least one objective.

Thus, to generate an efficient MOP, a model group can be formulated as follows:

Eff. $z(x) = [z1(x), ... zq(x)]$

based on

$x \in F$

where Eff. means seeking efficient solutions (in terms of minimizing and maximizing), while F represents the possible group, and x indicates the vector of decision variables. Therefore, based on this, in our case, it is first necessary to calculate the elements of the cost matrix. This matrix is the result of the optimization of each objective, here the costs of electricity generation and CO_2 emissions separately, giving the second goal the appropriate value for optimal solution of the first [51, 52]. Thus, a square matrix is obtained in which the level of conflict between the goals is reflected. Optimization is done with the application of the PHP-Simplex program [53, 54] from which we get the solutions. The application of this model in the research aims towards a more precise quantification comparison among the two conflicting aspects concerning electricity production – costs (operational costs) and the CO_2 emissions levels. This is especially interesting in the case of Kosovo, since the energy mix in the country is highly carbon-intensive because it originates predominantly from the domestic coal reserves.

2.6. Creating the Hierarchy of the Problem

In this analysis to define indicators for the decision-making and policy-making for introducing cleaner production technologies as a contribution to sustainable development, four main areas/main indicators are examined [55, 56, 57, 58, 59] which are modeled through the hierarchy of the problem presented in Figure 4. The four main groups presented in the first hierarchy level are:

- Environmental indicators
- Technical indicators
- Economic indicators
- Social indicators

The indicators that are identified in each of the groups, as well as the sub-indicators separately for each field, are shown in Figure 5. In the *second hierarchy level*, the following indicators with respect to each main group are presented:

- Environmental indicators
  – Resource efficiency

- CO_2 emissions
- Waste treatment
- PM emissions
- Soil contamination
- Landscape changes

- Technical indicators
  - Securing clean energy
  - Security of supply
  - Availability of energy
  - Availability of "know-how"
- Economic indicators
  - Costs
  - IRR – Internal Rate of Return
  - Decreasing of energy poverty (% of the household incomes for energy bills)

Energy poverty is a quite new term, without a precise and unique definition, but for the Western Balkan region the most appropriate definition is related to the decreasing number of properly heated households. That is, energy poverty is defined as a situation that households face when they are insufficiently heated, i.e., the average indoor temperature does not satisfy the requirement of thermal comfort [60], or when the heating level in the household is lower than a certain minimum that enables normal functioning in everyday activities [61].

- Social indicators
  - Health and safety
  - Good governance
  - Quality of life

The third and, in this case, the last hierarchy level in this problem's structure consists of the following sub-indicators:

- Environmental indicators
  - Energy efficiency
  - Efficiency of materials
  - CO_2 emissions per produced kWh

- Number of landfills
- Number of wastewater treatment plants
- PM10/PM2.5 concentration
- Number of contaminants/pollutants
- m^2 area/MW installed per different electricity generation technologies
- Technical indicators
  - Combined heat and power production
  - Use of biomass
  - Use of wind energy
  - Use of hydro power
  - Use of geothermal energy
  - Import dependency of energy commodities
  - Geopolitical issues
  - Natural disasters
  - Primary energy domestic reserves
  - Stochastic nature of the energy sources
  - Number of direct foreign investments
  - Number of foreign companies
  - Number of educated and skilled technicians
- Economic indicators
  - Investment costs (Euro/MW)
  - Operational costs
  - Costs related with the environment (externals)
- Social indicators
  - Air quality (average level of PM)
  - Number of premature deaths resulting from air pollution
  - Voice and accountability (VA)
  - Political stability and absence of violence (PV)
  - Government effectiveness (GE)
  - Regulatory quality (RQ)
  - Rule of law (RL)
  - Corruption control (CC)
  - GDP per capita
  - HDI – human development index

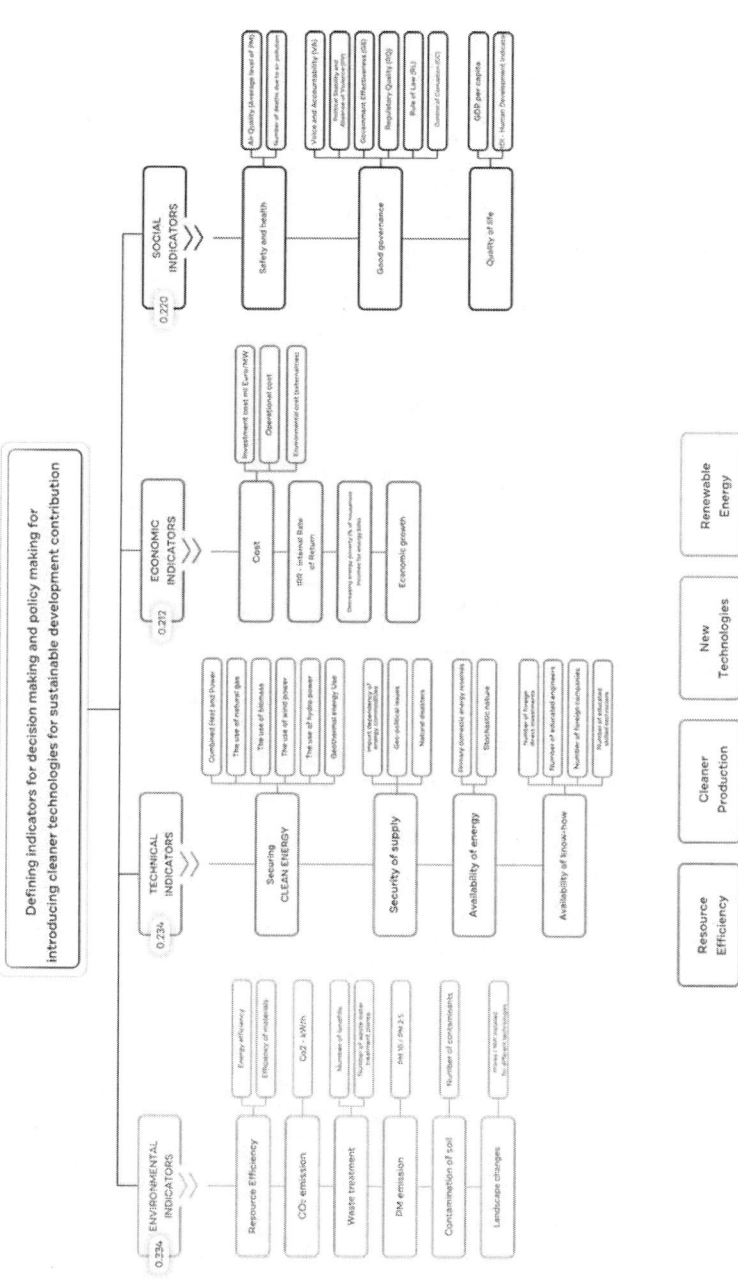

Figure 4. Hierarchy of the problem.

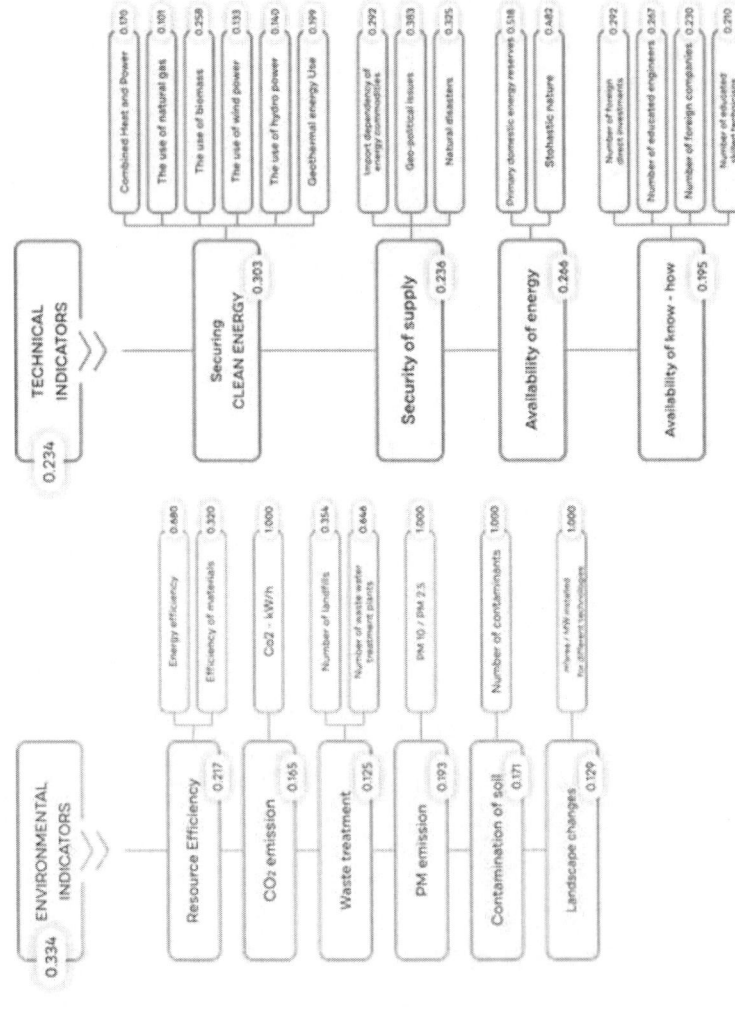

Figure 5. Indicators and sub-indicators separately for each field.

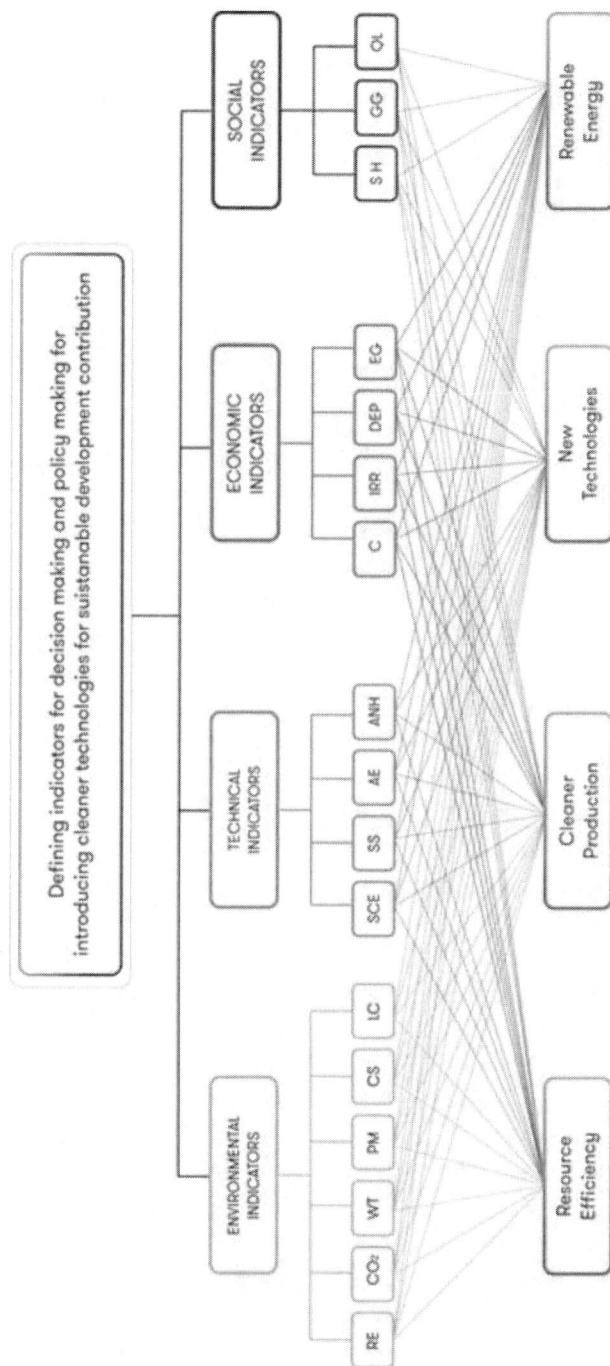

Figure 6. The comparative link between the second hierarchical level and alternatives.

The comparison of the pairs of indicators is performed through the entire hierarchy, ending with the third and final level. Further, the next phase of the research includes alternatives as a possible solution of the problem. The following four alternatives are chosen:

- Resource Efficiency
- Cleaner Production
- New Technologies
- Renewable Energy Sources

In Figure 6, there is a comparative link between the second hierarchical level and the alternatives. In this case, 16 indicators from the second level of the hierarchy will be compared with the four alternatives (Resource Efficiency, Cleaner Production, New Technologies, Renewable Energy). It should be emphasized that the total number of indicators on the second level is 17, but the Resource Efficiency indicator is omitted from the comparison since it has at the same time a role as one of the alternatives. The whole analysis is done based on the diagram shown in Figure 4. After analyzing the data and processing with Expert Choice software, we obtained the results.

The results will provide the ranking, which will be a good basis for the decision-making and policy-making for introducing policies that will contribute towards the sustainable development of the energy sector case study for Kosovo. Diagrams with the results obtained after processing all the data will be shown in the following chapters.

2.7. Hierarchy of the Problem Along with the Alternatives

The selection of the alternatives analyzed in this research focused on the situation in Kosovo has the goal of addressing the key segments that affect the energy sector's sustainable development, as well as the sustainable development of industry and the economy in general. Regarding the fact that Kosovo has significant coal reserves, the development towards the usage of domestic reserves must be oriented to the efficient use of resources (increased efficiency etc.), as well as the implementation of new technologies (supercritical parameters with high values of energy efficiency) for the new coal thermal power plants, in order to reduce the environmental impact. Of course, in this context, investments in the equipment required for reducing the emissions of sulfur and nitrogen oxides (externalities) are essential. This can be supported by the fact that the so-called energy transition that needs to be performed by the Western Balkan countries, on a long-term basis, requires energy from non-fossil sources. The sustainability of the overall energy system is, among others, assessed in terms of its negative environmental impact. This aspect is especially vulnerable in the Western Balkans region, where coal-fired thermal power plants still dominate the energy mix.

This research has the goal of developing a solid base for policy creators and decision-makers, particularly in the area of strategic paths of the energy generation infrastructure. Bearing in mind the measurable and quantified influences of the previously described indicators, they can be dealt with only by means of a comprehensive and integrated approach.

The introduction of the cleaner production concept arises as an additional benefit from these strategic policies, not only at the level of the energy sector's development, but more widely on the level of industry as a whole as well as the entire economy of the country. There is a great opportunity for SMEs, which are the most flexible and where the implementation of new technology could be a significant advantage for their competitive position. In this way, a synergy of the entire economic activity will be established at the SMEs, which will contribute to more rational energy consumption as well as a more responsible attitude towards environmental protection. In addition, this will have a positive impact on the stability and the sustainability of the energy sector.

Chapter 3

Concept and Application of Resource Efficient and Cleaner Production (RECP)

As far as the RECP analysis is concerned, the basic meaning of Resource Efficient and Cleaner Production should be clarified.

Resource Efficient and Cleaner Production (RECP) has come to refer to preventive environmental measures to facilitate pollution prevention and reduce the carbon intensity per unit of products, along with the financial profit of the industry. Energy Efficiency (EE) and Cleaner Production (CP) measures in industry can facilitate the promotion of low carbon industrial development through pollution prevention and energy conservation [62, 63].

Since the 1990s, the United Nations Environment Program (UNEP) and the United Nations Industrial Development Organization (UNIDO) have collaborated in order to foster and promote sustainable industrial production through the Resource Efficient and Cleaner Production (RECP) Program in developing countries and economies in transition [64]. RECP means the integrated and continuous application of preventive environmental strategies to processes, products, and services to increase efficiency and reduce risks to humans and the environment [65]. Such an approach is presented in Figure 7.

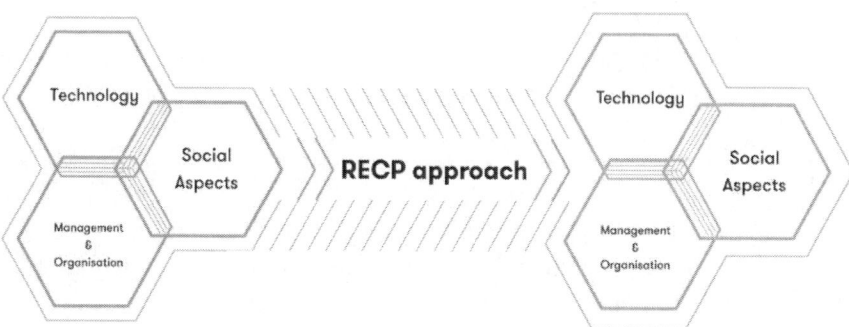

Figure 7. Resource Efficient and Cleaner Production as an integrative approach to develop sustainable companies (Fegerl, 2016).

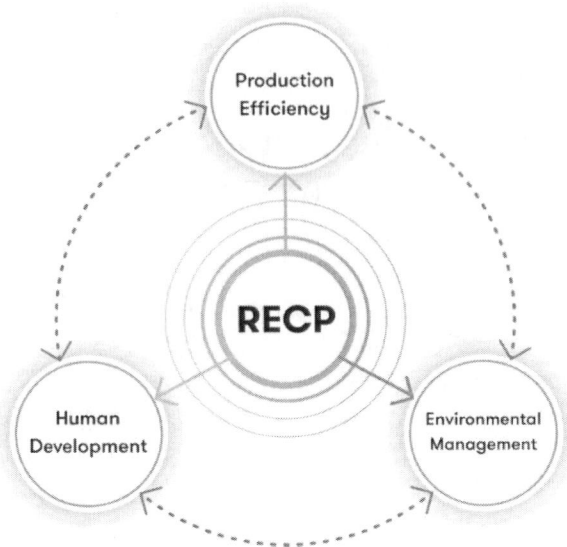

Figure 8. Three dimensions of sustainable development in an integrated manner.

RECP addresses the three dimensions of sustainable development individually and synergistically through a systematic approach (UNEP, 2016) [66]. Figure 8 presents the three dimensions of sustainable development in the integrated RECP approach.

- Technology - Production efficiency: optimization of productive use of natural resources (materials, energy and water)
- Organization - Environmental management: minimization of impacts on environment and nature
- Social - Human development: minimization of risks to people and communities and empowering their sustainable development

The RECP approach is built on good housekeeping and suggests mapping energy and resources, waste, water and emissions to their respective sources in the process, to analyze the reasons for their generation and to identify options as close as possible to the source in order to minimize the generation of waste and emissions. These principles are congruent with the cyclic principles of a circular economy [67, 68].

Cleaner production, on the contrary, aims to reduce both the negative effects on the environment and the operating costs. Cleaner production works with process-integrated – preventive – methods instead of "end-of-pipe"

Concept and Application of Resource Efficient and Cleaner Production

solutions. Cleaner production is the conceptual and procedural approach to production that demands that all phases of the life cycle of a product or of a process should be addressed with the objective of the prevention or minimization of short- and long-term risks to humans and to the environment [69]. Figure 4 presents the components of cleaner production as a strategy that has a preventive, continuous and integrative role in products and processes, aiming towards the reduction of the risks for people and the environment.

Cleaner production (CP) has great importance in the field of environmental policy and management. Environmentally, the CP approach provides a concrete and long-term technique to eliminate and/or reduce such emissions as carbon dioxide and sulfur dioxide. Consequently, CP plays important roles also in addressing global environmental issues such as climate change, acid precipitation, and urban smog [70, 74].

Five basic principles of cleaner production are required for efficient resources management. They consist of the careful use of resources, the closing of material streams, and resource substitution. Table 4 explains the five main principles of cleaner production, which encompass all phases of the product. It must be emphasized that such an approach treats as equally important the inputs (raw materials) and the procedures after the expiry of the product's life. In this context, the manner of raw materials production (their impact on the environment), as well as the waste collection, recycling etc., have the same importance as good management in terms of the rational and efficient use of resources, cost optimization and introduction of new technologies.

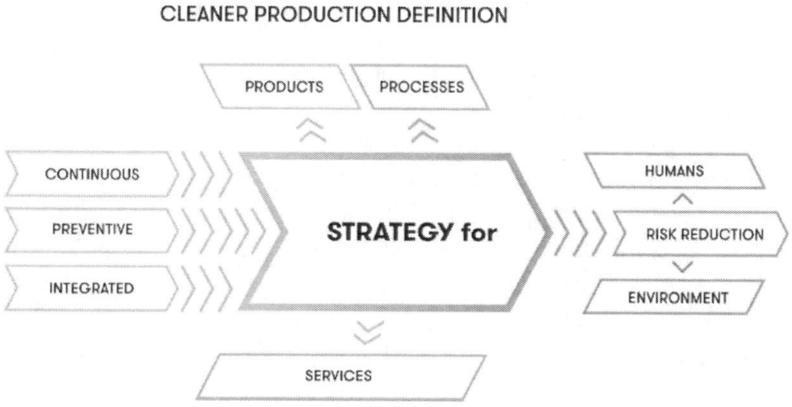

CLEANER PRODUCTION DEFINITION

PRODUCTS / PROCESSES

CONTINUOUS →
PREVENTIVE → **STRATEGY for** → HUMANS / RISK REDUCTION / ENVIRONMENT
INTEGRATED →

SERVICES

Figure 9. Cleaner production definition.

Table 4. Cleaner production technologies and tools for Resource Efficient production [72]

Input-Substitution
Use of less hazardous raw, auxiliary or operating materials.
Use of operating materials with a longer lifetime.
Good Housekeeping
Increase the material and energy efficiency of actions in the process.
Try to deal with the "low hanging fruits" first, e.g. reduce losses due to leakage.
It is important to train employees.
Internal Recycling
Close material and energy loops for water, solvents, etc.
Cascading of material and energy streams.
Technological Optimization/Change
Implementation of new technologies.
Improved process control.
Redesign of processes.
Change in or substitution of hazardous processes.
Optimization of the Product
Increasing the lifetime.
Easier repair.
Easier de-manufacturing, recycling or disposal.
Use of non-hazardous materials.

3.1. The Concept and Indicators by UNIDO

The system that is herein suggested consists of six absolute indicators - three in terms of the use of resources (energy, materials and water) and three concerning pollution (air emissions, waste water and waste), as well as one reference indicator – the specific product. These absolute indicators are used in turn to determine three indicators concerning the resource's productivity (products per unit energy consumption) and three for pollution (air emissions, wastewater and waste). The indicators have been selected on the basis that they collectively cover the most important environmental aspects of SME operation and that improvements in these areas generally provide the maximum benefits for the environment and the business. Moreover, the data required for these absolute indicators should, at least in principle, be available or measurable by any company, resulting in a relatively low implementation cost in relation to the potential benefits [73].

Figure 10. Five basic principles of cleaner production (CP).

The indicators referring to resource use are:

Energy use: final energy use of the company, measured in mega joules or kilowatt hours, including the energy content of fuels used (gas, oil, petrol, biomass, etc.) and electricity consumption.
Materials use: total mass of materials used by the company, measured in tons, including raw materials, packaging and distribution materials, auxiliary materials, etc., but excluding the weight of fuels.
Water use: total water consumption of the company, measured in kiloliters or cubic meters, including all sources (ground water, tap/drinking water, surface water) and all applications (process water, cooling water, sanitary water, etc.).

Based on the five principles and the corresponding information presented in Table 1, the diagram presented in Figure 10 is prepared.

Chapter 4

Defining Indicators for Decision and Policy Making for Contributing to Sustainable Development

4.1. Making Decisions

As stated in the previous chapters, this analysis focuses on defining the indicators of decision- and policy-making for sustainable development of the electricity generation structure in Kosovo, as well as the efficient use of energy in SMEs. The results obtained in this regard have to do with how decision-making is defined, which in our case is done using the software - Expert Choice (EC) [74].

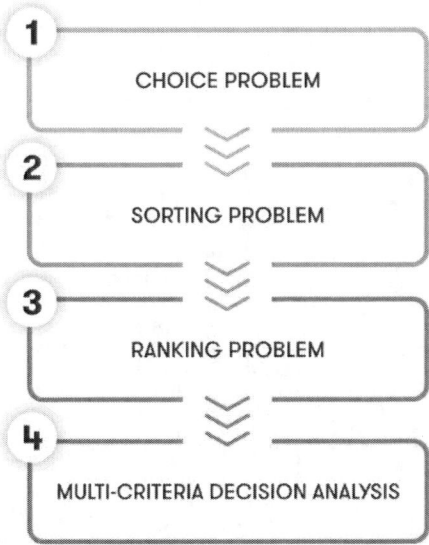

Figure 11. Four main types of decision.

The problems of decision-making are often hard to solve, and when they touch the multi-criteria domain, they become even more complicated. But what is, in fact, defined as a decision problem? In their everyday lives, people face various decision-making situations. Those decisions are different, based on the criteria used. Nevertheless, the division proposed by Roy (1981) [75], who identified four different types of decisions (Figure 11), is the one that has been taken into account here. In this model, note Position 4, Multi-Criteria Decision Analysis.

Multi-criteria decision-making can be defined as the process of evaluating situations from the real world based on qualitative/quantitative criteria in a certain/uncertain/risky environment, the ultimate goal of which is to propose the best directions of action/activities/choice/strategy/policy between available options, [36, 76].

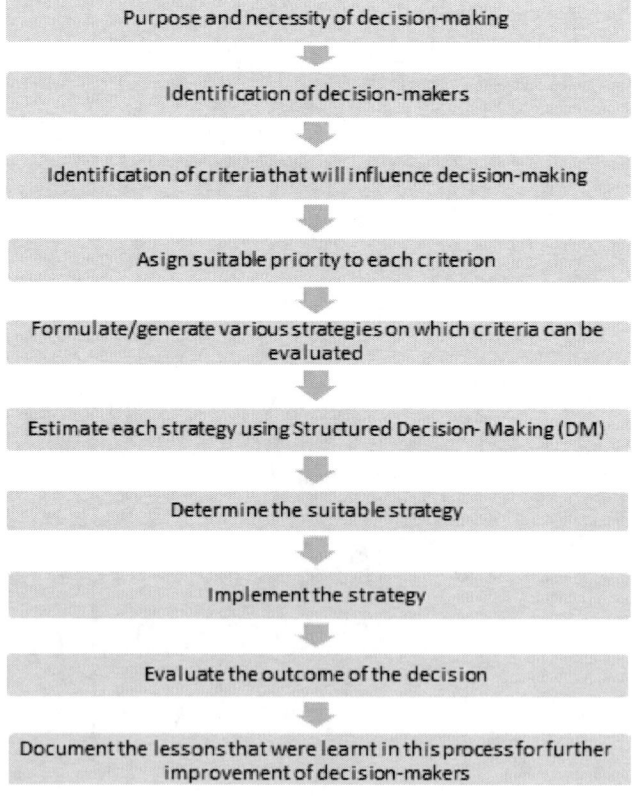

Figure 12. The methodological steps in the structured Multi-Criteria Decision.

Figure 12 presents the methodological steps in the structured multi-criteria decision.

1. *Choice problem.* The goal is to choose the best solution, or to reduce the number of possible options in the sub-group of the equivalent or incomparable "good" options. For example, a manager chooses the appropriate person for a certain project.
2. *Sorting problem.* The options are listed in predefined groups or categories. The task is to regroup the options with similar behavior or with similar features for the descriptive, organizational and predictable causes. For example, employees could be categorized as extraordinary, average and weak workers. Based on these classifications, the required measures could be undertaken.
3. *Ranking problem.* The solutions are listed from the best towards the worst, including parallel comparisons etc. The ranking could even be partial when the options are considered to be incomplete or incomparable. A typical example is the ranking of universities based on several criteria such as the quality of the education process, research work and the career possibilities.
4. *Multi-Criteria Decision Analysis. Descriptive problem.* The task is to describe the solutions and their consequences. This is usually performed as a first step in understanding the features of the analyzed problem [77, 78].

4.2. Pyramid of Information, Simple and Complex Indicators

Indicators can contain simple and significantly aggregated information. Figure 13 displays a pyramid of information which basically begins with rough (raw) data, processed further into statistical data, simple indicators and finally into complex indicators. The level of aggregation depends on the requirements of the user or the purpose of the research.

In essence, different rules, aspects and approaches move the world of scientists, the world of policy-makers and politicians, and the world of economics and industry, i.e.,

- Scientists concentrate on detail, confidentiality, accuracy, etc.;
- Economic experts focus on the dimension of the potential profit that might be expected from planned investments in industry and the economy;
- High-level politicians are interested in the wider picture, for the key message, preferably a value accumulated in one digit on a scale from 0 to 10 [79].

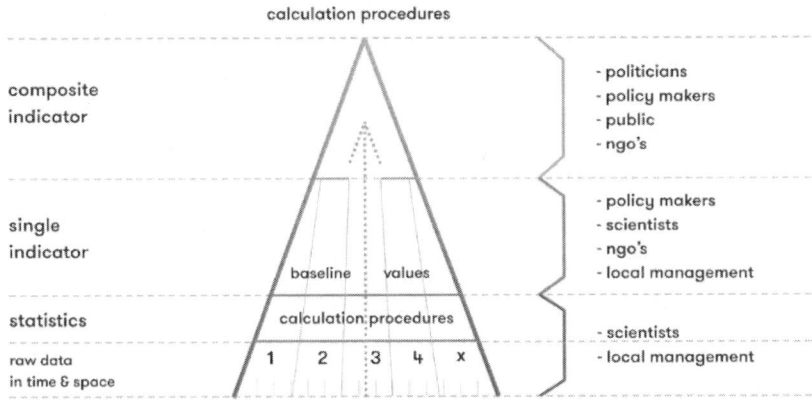

Figure 13. Information pyramid. Source: Ben Ten Brink, 2006.

4.3. Expert Choice Software

It has been emphasized that the decision-making process requires accurate and careful treatment of the problems. Software developments and their application have enabled decision-making to become steadily better, more competent and to have consistency in the process. Of course, in the framework of multi-criteria decision-making there are several methods and different software programs have been developed that can be most effectively applied to analyze different problems. Table 5 presents these methods and the decision problems they solve.

AHP can also be considered as a mechanism that offers great help to decision-makers by providing them with well-organized structures to determine the importance of objective evaluations to the decision-making by offering alternatives. The theoretical AHP Method Theory, which is implemented easily, quickly and precisely by the Expert Choice software, provides optimum solutions for even more complicated problems requiring multi-criteria treatment.

Table 5. Methods and the decision problems they solve [36]

MCDA SOFTWARE PROGRAMS		
Problem	MCDA Method	Software
Ranking, description, choice	PROMETHEE – GAIA	Decision Lab, D-Sight, Smart Picker Pro, Visual Promethee
Ranking, choice	PROMETHEE	DECERNS
	ELECTRE	Electre IS, Electre III – IV
	UTA	Right Choice, UTA +, DECERNS
	AHP	EXPERT CHOICE, MakeItRational, Decision Lens, HIPRE 3+, RightChoiceDSS, Criterium, EasyMind, Questfox, ChoiceResults, 123AHP, DECERNS
	ANP	Super Decisions, Decision Lens
	MACBETH	M-Macbeth
	TOPSIS	DECERNS
	DEA	Win4DEAP, Efficiency Measurement System, DEA Solver Online, DEAFrontier, DEA-Solver PRO, Frontier Analyst
Choice	Goal Programming	-
Sorting, description	FlowSort – FS-GAIA	Smart Picker Pro
Sorting	ELECTRE – Tri	Electre Tri, IRIS
	UTADIS	-
	AHPSort	-

The Expert Choice (EC) software is a multi-objective decision support tool based on the Analytic Hierarchy Process (AHP), a mathematical theory first developed at the Wharton School of the University of Pennsylvania by one of Expert Choice's founders, Thomas Saaty (1977) [80, 81].

The following steps are used in AHP and Expert Choice:

1. Brainstorm and structure a decision problem as a hierarchical model
2. Set the type and mode of pairwise comparisons or data grid functions
3. Group enable the model
4. Import data to Expert Choice from external databases
5. If applicable, compare pairwise the alternatives for their preference with respect to the objectives, or assess them using one of the following: ratings or step functions, utility curves, or entering priorities directly

6. Compare pairwise the objectives and sub-objectives for their importance to the decision
7. Synthesize to determine the best alternative
8. Perform sensitivity analysis
9. Export data to external databases
10. Perform resource allocations using Expert Choice's 'Resource Aligner' to optimize alternative projects subject to budgetary and other constraints.

The steps used in AHP with EC are shown in Figure 14 below:

- Brainstorming and setting up the hierarchical structure
- Comparison in pairs of criteria and sub-criteria in relation to their importance in making the decision
- Comparison in pairs of alternatives and determining preferences in relation to criteria or their assessment

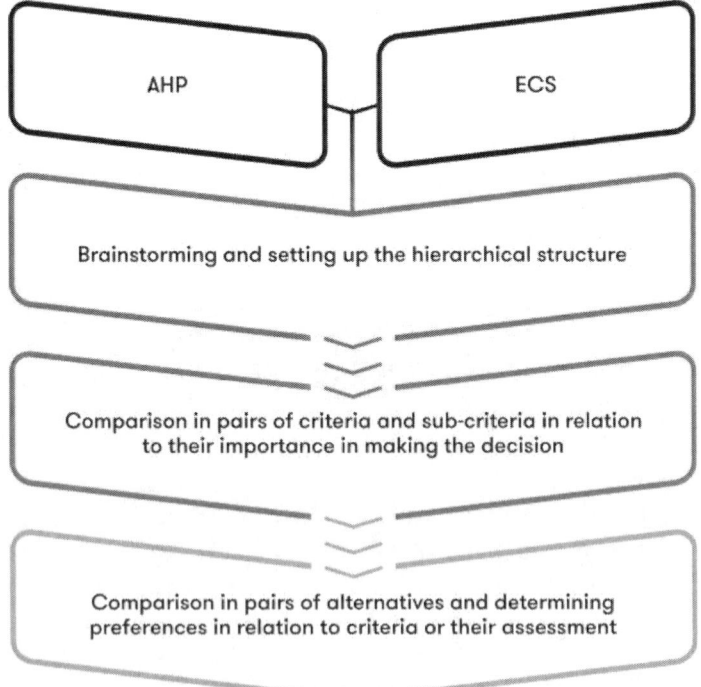

Figure 14. Steps used in AHP with the EC software.

Defining Indicators for Decision and Policy Making ... 43

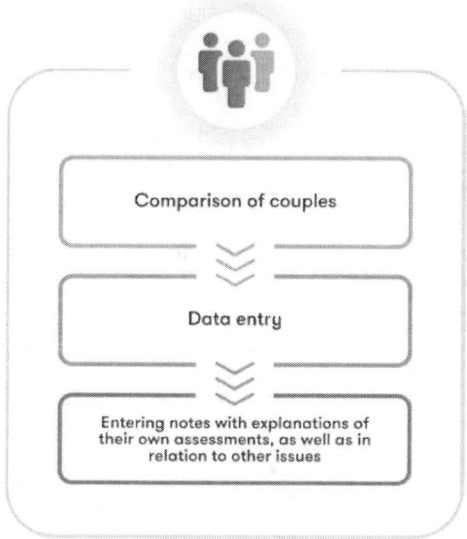

Figure 15. Participants can make evaluations.

EC can also be used for teamwork in order to improve the quality of group decision-making and to synthesize different opinions. Many decisions are really complex, and one decision-maker cannot properly synthesize all the relevant information in the decision-making process. Accordingly, the group has the potential to generate more ideas and of course knows more than the individual [82, 83]. The facilitator builds the model and coordinates the decision-making process of the group. All members of the group or team are marked as "participants."

The participants can make evaluations on various aspects of the problem that is the subject of decision-making, including (Figure 15).

- Comparison of couples
- Data entry
- Entering notes with explanations of their own assessments, as well as in relation to other issues.

Participants can see their own decision-making model, perform synthesis, and review the content of the so-called Data Grid. All information can be printed. If it is allowed by the facilitator, participants may have insight and can print one or more of the combined (summary) results [84]. Web models of the EC allow group members from different parts of the world to simultaneously open up the same model to solve problems and decision-making [85].

4.3.1. List of Experts/Participants in the Evaluation Process and Questionnaire

The way to achieve real results using EC software, however, requires the not easy process of completing the cycle of filling out the surveys. When we talk about decision-making in the domain of cleaner production and resource efficiency, then receiving reliable responses necessarily requires the contributions of professionals in the field. Our analysis is based on four major areas of special importance, for which we have chosen experts with credible professional and scientific backgrounds. In Table 6 we present the profiles and institutions of the experts that have contributed consistent opinions regarding our analysis.

Determining weight factors is among the most important points for defining the problem correctly through the usage of AHP. AHP makes it possible to compare couples and determine their importance/preference, which is one of the great advantages of this method. Through this, we can also identify useless/inconsistent participants and those who do not take a serious approach to our analysis. In cases where participants in the analysis need to address a large number of comparisons, it is very important that they are willing to concentrate on the whole problem trajectory, thus giving consistent answers and opinions. In our case, the participants could choose the form of their responses, some of them through direct communication and others by completing the questionnaire electronically, or sending their contributions via email. The entire questionnaire (Appendix 1) was compiled based on the preferences that are determined by Saaty's ranking (Figure 16).

	Extremely		Very strong		Strong		Moderately		Equal		Moderately		Strong		Very strong		Extremely	
Criteria A	9	8	7	6	5	4	3	2	1	2	3	4	5	6	7	8	9	Criteria B

Figure 16. Example of an indirect determination of weight factors by pairwise comparison (AHP) by making a direct interview questionnaire.

Table 6. Profiles and expert institutions that contributed to the research

Nr	Institution	Position
1,2,3	University of Pristina Mechanical Engineering Faculty	Professors
4	University of Skopje Mechanical Engineering Faculty	Professor
5	University of Westminster, UK	Master of Economic Policy and Data Analysis
6	University of Tetova, N. Macedonia	Professor
7	JSC Macedonian Power Plants	PhD in technical sciences, senior engineer for process analyses
8,9,10	University of Mitrovica Faculty of Mechanical and Computer Engineering	Professor
11	University of Pristina Faculty of Philosophy	Sociologist
12	Mayor of Municipality	Mayor
13,14	Kosovo Energy Corporation	Engineers
15	District Heating "Termokos" Department of Distribution	Engineer
16	District Heating "Termokos" Department of Production	Engineer
17	District Heating "Termokos" Member of PIU Project Implantation Unit – Cogeneration project	Engineer
18	Ministry of Economic Development	Head of Department
19	Ministry of Transport	Head of Department
20	Ministry of Environmental and Spatial Planning	Head of Department
21	Ministry of Finance	Head of Department
22	Regulatory Office of Energy in Kosovo Member of Board	Board Director
23	University of Pristina Faculty of Mathematics and Natural Sciences – Chemistry Department	MSc in Analytical and Environmental Chemistry
24	Kosovo Energy Efficiency & Renewable Energy Project Ministry of Economic Development	Engineer
25	EFACEC Contracting Central Europe GmbH Master in Business Administration	Regional Director

The figure shows the scaling of weight factors by pairwise comparison (AHP) as the basis of the EC software used to achieve results. This procedure is shown in Figure 16.1.

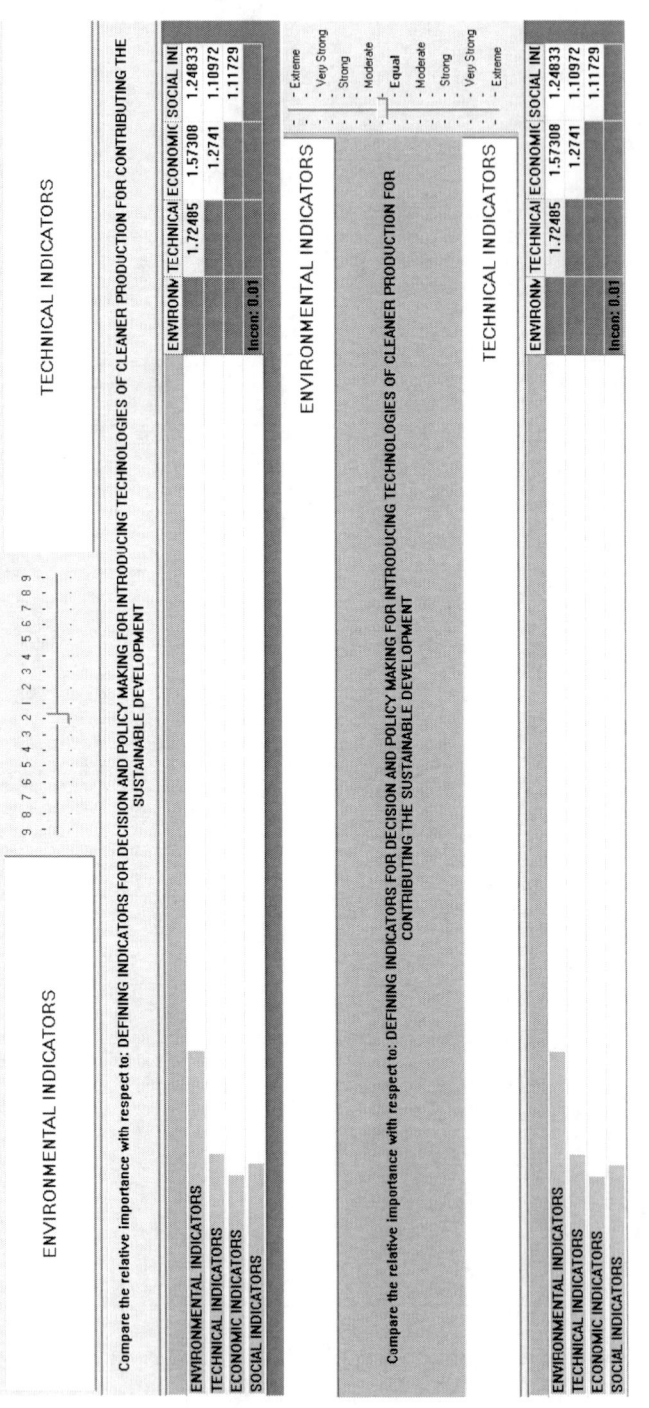

Figure 16.1. (Continued)

Defining Indicators for Decision and Policy Making …

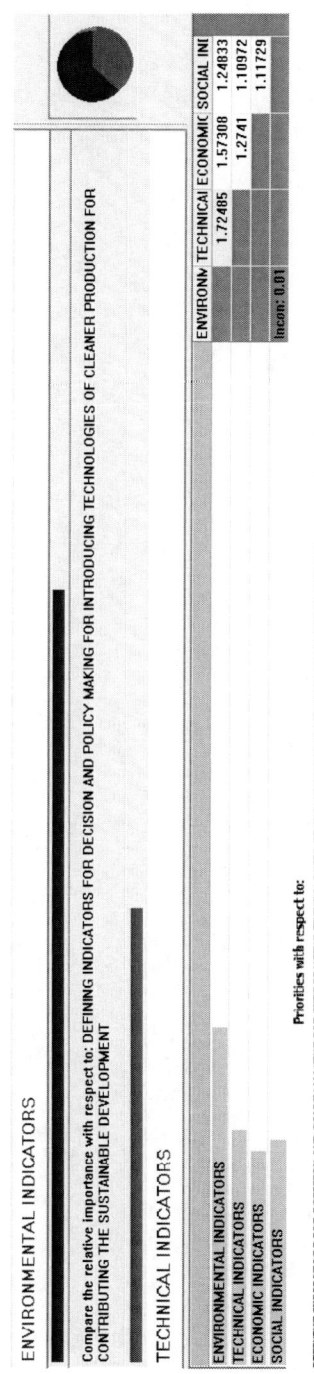

Figure 16.1. Results achieved through Expert Choice during analysis of weight factors by pairwise comparison. The analysis shows the comparison of Environmental Indicators and Technical Indicators.

Chapter 5

Applied Methodology: Process and Method of Multi-Criteria Decision-Making (MCDM), Analytic Hierarchy Process (AHP) and Multi-Objective Programming (MP)

5.1. Calculation of the Weight Factors of the Indicators

To calculate the weight factors of each of the indicators, the AHP method is used as previously explained. Using the EC software, the weight factor of the indicators from each participant in the test is calculated, i.e., aggregation of the weight factors of all participants, corresponding to the elements of the matrix A_i [86]:

$$A_i = \begin{bmatrix} 1 & \cdots & a_{1n_i} \\ \cdots & 1 & \cdots \\ 1/a_{1n_i} & \cdots & 1 \end{bmatrix}, i = 0, l-1, l = \sum_L l_L \tag{18}$$

where L represents the number of hierarchical levels. The sum of the determined weight factors from each hierarchical level should correspond to the next expression:

$$\sum_{j=1}^{n_i} w_j = 1 \tag{19}$$

In addition, the software also performs consistency checks to exclude inconsistent responses or participants from the analysis.

5.1.1. Determination of Alternatives and Their Ranking

In order to analyze/calculate the real/suitable/potential options to make decisions about introducing technologies of cleaner production for contributing to sustainable development, we decided to analyze four alternatives.

Alternative 1 - Resource Efficiency (A1)
Alternative 2 - Cleaner Production (A2)
Alternative 3 - New Technologies, (A3)
Alternative 4 - Renewable Energy, (A4)

After analyzing the entire model, and thus all levels in the hierarchy of our problem, the results are obtained and are presented in Figure 17. Since the alternatives from which the results are derived are obtained from the complete model analysis, in this is set on the four alternatives to only two levels of the problem hierarchy. So, the alternatives are analyzed by taking the hierarchy of the two levels, in respect to all of the indicators: environmental, technical, economic and social. The software is the same as that used for achieving the results of the whole model, and it is also used for the ranking of alternatives. Thus, the multi-criteria program/software takes into account the predetermined factors of the weight indicators.

In Figure 18, the hierarchy of the problem appears under the logic of the analysis of alternatives. If the results obtained are analyzed in terms of the alternatives, then the ranking given in Table 7 emerges.

Table 7. Ranking of alternatives

Alternative 1	Cleaner Production	0.304	1
Alternative 2	Resource Efficiency	0.301	2
Alternative 3	Renewable Energy	0.295	3
Alternative 4	New Technologies	0.281	4

The results in Table 7 show that Alternative 1 - Cleaner Production has the highest importance in relation to the other three alternatives. Therefore, it is necessary to work continuously on creating conditions and providing investment in Cleaner Production. Resource Efficiency, Renewable Energy and New Technologies are also of extraordinary importance as alternatives, as derived from the analysis of the whole model, in which the indicators for sustainable development are defined. From this point of view, sustainable development tends with a slight advantage from Alternative 1, but it requires permanent ignition with the steps to be taken in relation to Alternatives 2, 3 and 4. The results of the alternatives are given through the visual representations shown in Figures 20–38, while the results obtained for each of the sixteen indicators are given in Figures 18 and 19, and Tables 10, 11, 12 and 13, which represent the values of indicators from the second level, as compared with the four alternatives.

The definition of the indicators in the four major areas that are taken as the basis for this analysis is based on the basic pillars of sustainable development (Environmental Indicators, Technical Indicators, Economic Indicators and Social Indicators) (Figure 18). The creation of the hierarchy model is based on the main Goal – starting at the first level with 4 main indicators, which in the second level are expanded with 17 other indicators, completing the third level of the hierarchy of the model with 36 indicators. The total hierarchy structure has 57 indicators, the results of which will certainly give a good picture of the analysis of sustainable development in the Republic of Kosovo, and so also for other countries in the region. The results obtained after processing in Expert Choice software are presented in Figures 39–49.

The results in terms of the Environmental Indicators show that, of the values at the second level hierarchy, Resource Efficiency (0.217) was the highest, while at the third level the highest were Energy Efficiency (0.680) and the Number of wastewater treatment plants (0.646).

Results for the Technical Indicators show that, of the values at the second level of the hierarchy, Securing Clean Energy (0.303) was the highest, while at the third level of the hierarchy the Use of biomass (0.258), Geo-political issues (0.383), Primary domestic energy reserves (0.518), and Number of foreign direct investments (0.292) had the highest values.

For the Economic Indicators the results show that, at the second level of the hierarchy, Economic Growth (0.412) had the highest value, while at the third level the Environmental Cost (externalities) (0.415) had the highest value.

The results for Social Indicators show that, at the second level of the hierarchy, Safety and Health (0.452) had the highest value, while at the third level Air Quality (average level of PM) (0.588), Control of Corruption (0.519) and GDP per capita (0.519) had the highest values.

In this analysis, the ranking of the alternatives can be performed for two cases – the real values of the weighting factors as per the experts' preferences, and an assumed (academic) case of equal values of weighting factors [59]. In this book, the latter is not performed. Hence, the total number of indicators in relation to which the ranking is conducted is 16 (according to Figure 18).

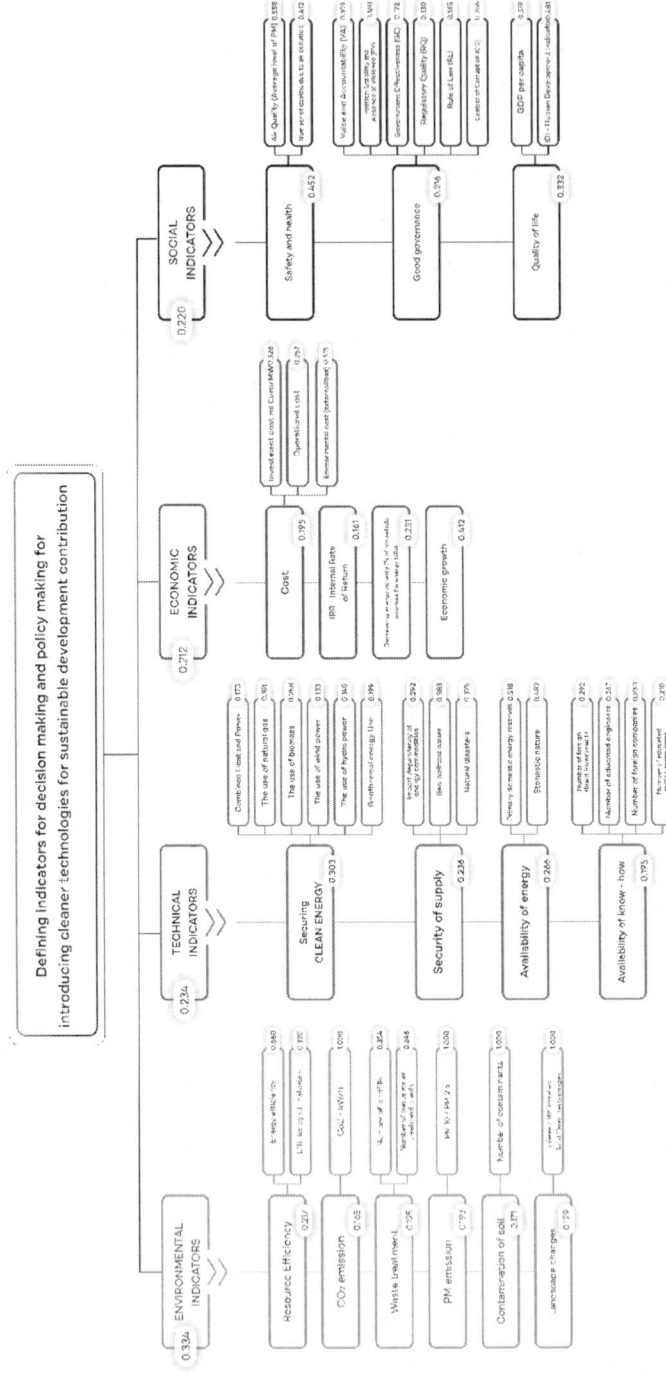

Figure 17. Diagram of problem hierarchy. Presentation of all indicators for environmental, technical, economic and social indicators, from goal to final levels.

Table 8. Evaluation of alternatives to indicators in the case of equal weight factors

Analyzed alternatives	Environmental Indicators 0.334						Technical Indicators 0.234				Economic Indicators 0.212				Social Indicators 0.220		
	CO2 Emission	Waste Treatment	PM Emission	Contamination of Soil	Landscape Changes	Securing Clean Energy	Security of supply	Availability of energy	Availability of knowhow	Cost	IRR – Internal Rate of Return	Decreasing energy poverty (% of household incomes for energy bills)	Economic growth	Safety and health	Good Governance	Quality of life
	0.165	0.125	0.193	0.171	0.129	0.303	0.236	0.266	0.195	0.195	0.161	0.231	0.412	0.425	0.216	0.332
A1	0.15	0.35	0.35	0.35	0.35	0.15	0.35	0.35	0.35	0.35	0.35	0.15	0.35	0.35	0.35	0.35
A2	0.35	0.35	0.35	0.35	0.35	0.35	0.35	0.15	0.35	0.15	0.35	0.35	0.35	0.35	0.35	0.35
A3	0.35	0.35	0.15	0.35	0.35	0.35	0.15	0.15	0.35	0.35	0.15	0.15	0.15	0.35	0.35	0.35
A4	0.35	0.15	0.35	0.35	0.35	0.35	0.15	0.35	0.35	0.35	0.35	0.15	0.35	0.35	0.35	0.35

Figure 18. The results obtained on the alternatives offered by our mode.

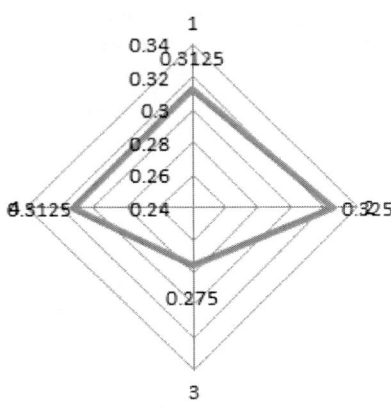

Figure 19. Final/summarized results of alternatives. Analysis of environmental, technical, economic and social indicators.

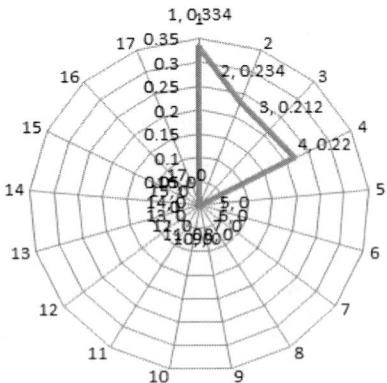

Figure 20. Final results from the analysis of the environmental, technical, economic and social indicators, for the first level of the hierarchy of our model.

There are previously determined values for the evaluation of each of the alternatives with respect to each indicator, with values on the scale from 1 to 10 as follows [59]:

Best Score 10 (0.5)
Average rating 7 (0.35)
Low rating 3 (0.15)

The grading values are the appropriate interpretations of the scale scores offered by the software.

The total value of the grades is 1. The decision-making software includes appropriate ratings for each alternative, based on relevant literature data as well as a personal assessment (based on relevant literature). These data are presented in Table 8. Then the inscribed grades are normalized, that is, their sum for each column (the four alternatives) should be a unit. The software calculates the weights of each alternative, taking into account not only the assessments in the specific case described but also the corresponding predetermined (equal) weight factors of the 16 indicators under consideration.

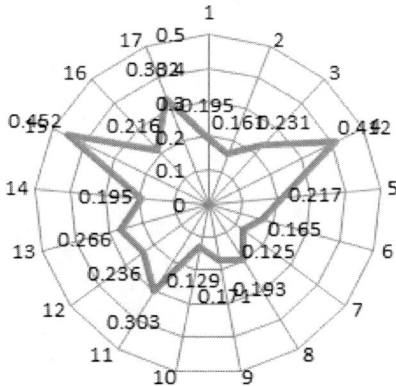

Figure 21. Final results from the analysis of the environmental, technical, economic and social indicators, for the second level of the hierarchy of our model.

The complete results of the analysis of all indicators – the part of the model for the three levels of the hierarchy together with the results obtained by the Expert Choice software – are presented in Figures 38– 49.

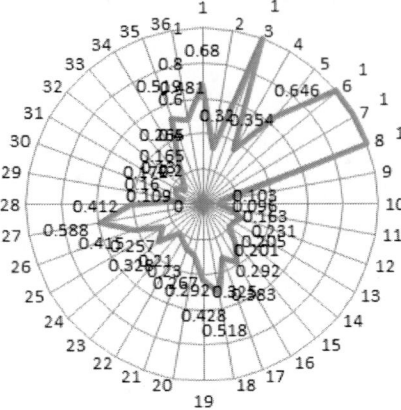

Figure 22. Final results from the analysis of the environmental, technical, economic and social indicators, for the third level of the hierarchy of our model.

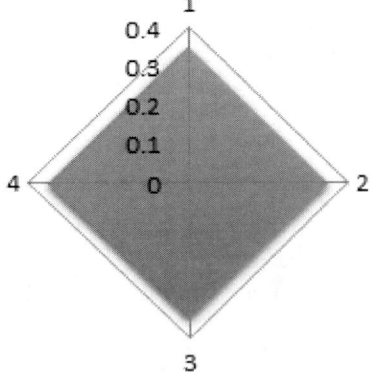

Figure 23. Introducing the analyzed values between the second hierarchical level and the alternatives – Safety and Health/Social Indicators.

Applied Methodology

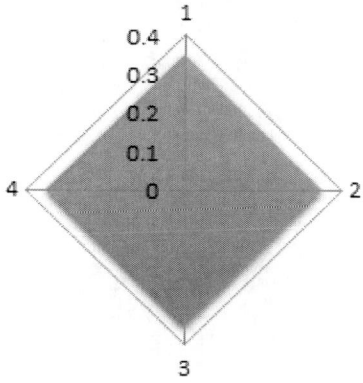

Figure 24. Introducing the analyzed values between the second hierarchical level and the alternatives – Good Governance/Social Indicators.

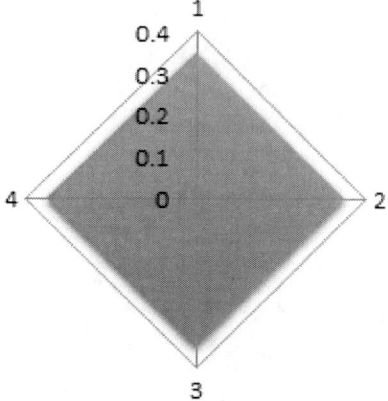

Figure 25. Introducing the analyzed values between the second hierarchical level and the alternatives – Quality of Life/Social Indicators.

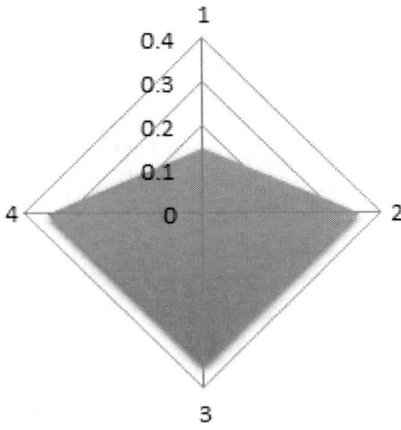

Figure 26. Introducing the analyzed values between the second hierarchical level and the alternatives – CO_2 emission/Environmental Indicators.

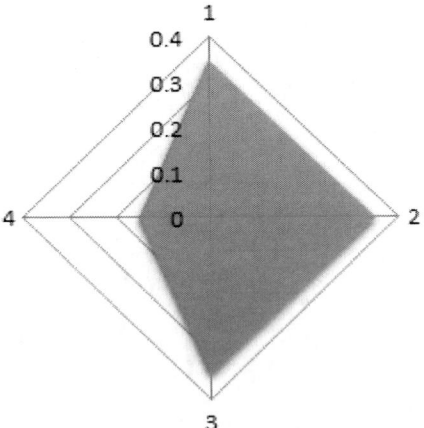

Figure 27. Introducing the analyzed values between the second hierarchical level and the alternatives – Waste Treatment/Environmental Indicators.

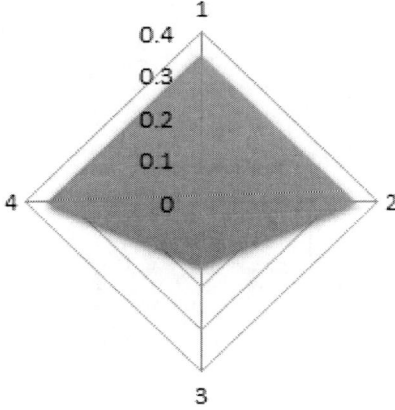

Figure 28. Introducing the analyzed values between the second hierarchical level and the alternatives – PM Emission/Environmental Indicators.

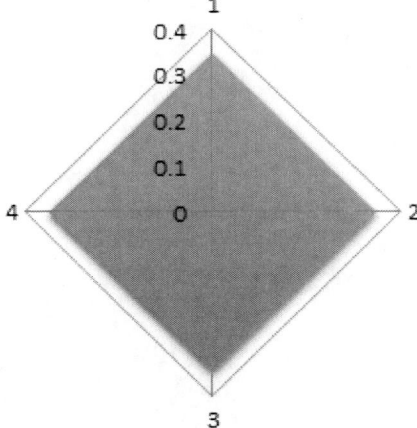

Figure 29. Introducing the analyzed values between the second hierarchical level and the alternatives – Contamination of Soil/Environmental Indicators.

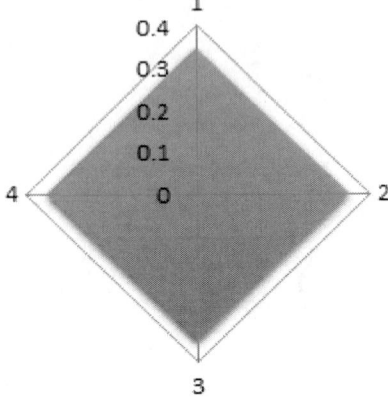

Figure 30. Introducing the analyzed values between the second hierarchical level and the alternatives – Landscape Changes/Environmental Indicators.

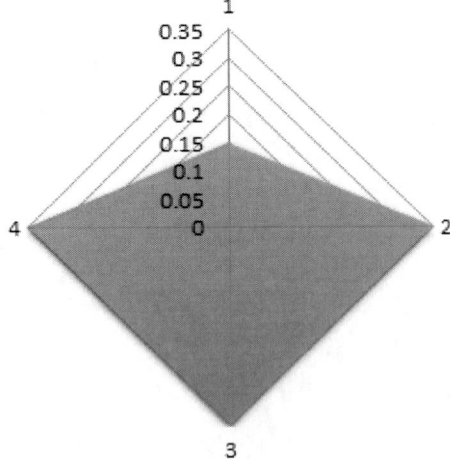

Figure 31. Introducing the analyzed values between the second hierarchical level and the alternatives – "Securing Clean Energy"/Technical Indicators.

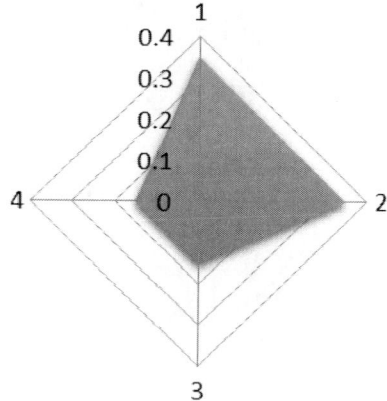

Figure 32. Introducing the analyzed values between the second hierarchical level and the alternatives – Security of Supply/Technical Indicators.

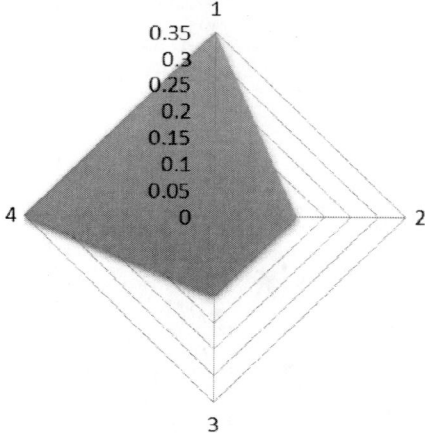

Figure 33. Introducing the analyzed values between the second hierarchical level and the alternatives – Availability of Energy/Technical Indicators.

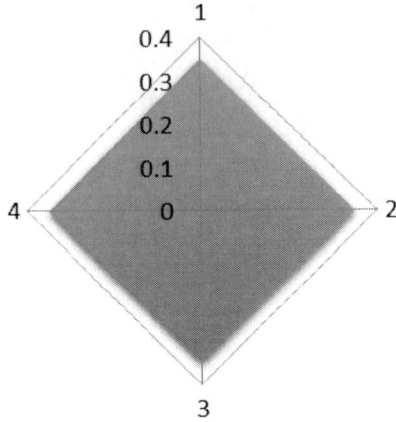

Figure 34. Introducing the analyzed values between the second hierarchical level and the alternatives – Availability of Know-How/Technical Indicators.

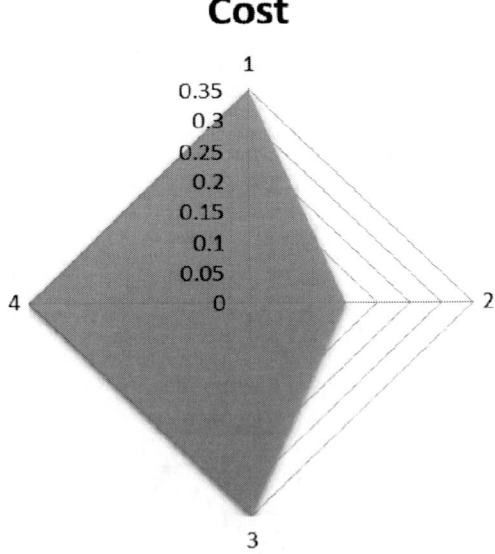

Figure 35. Introducing the analyzed values between the second hierarchical level and the alternatives – Cost/Economic Indicators.

Applied Methodology 63

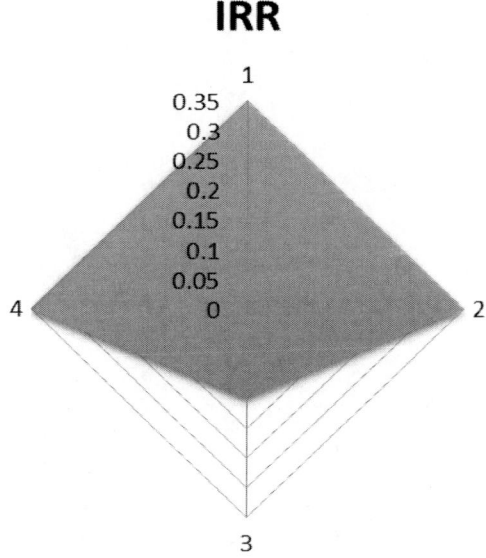

Figure 36. Introducing the analyzed values between the second hierarchical level and the alternatives – IRR/Economic Indicators.

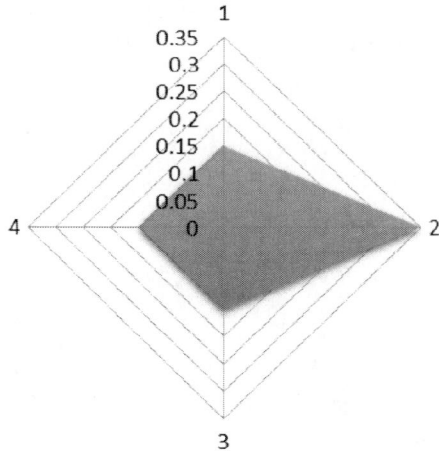

Figure 37. Introducing the analyzed values between the second hierarchical level and the alternatives – Decreasing Energy Poverty/Economic Indicators.

Environmental Indicators

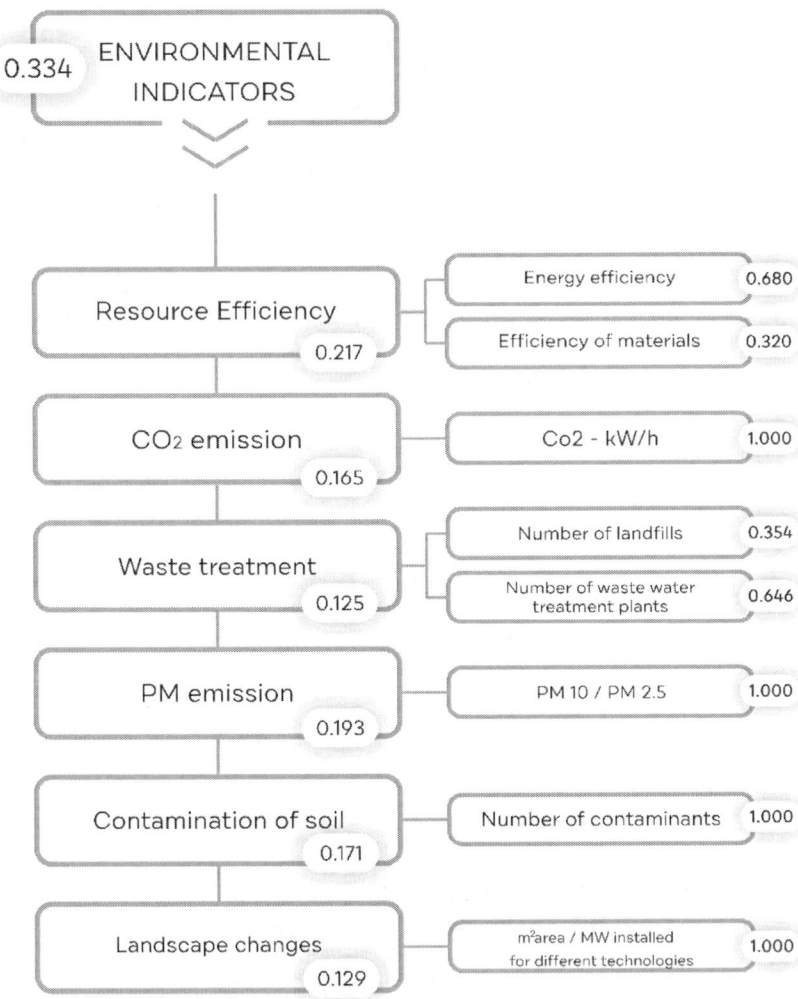

Figure 38. Graphical representation of results for all hierarchical levels for Environmental Indicators.

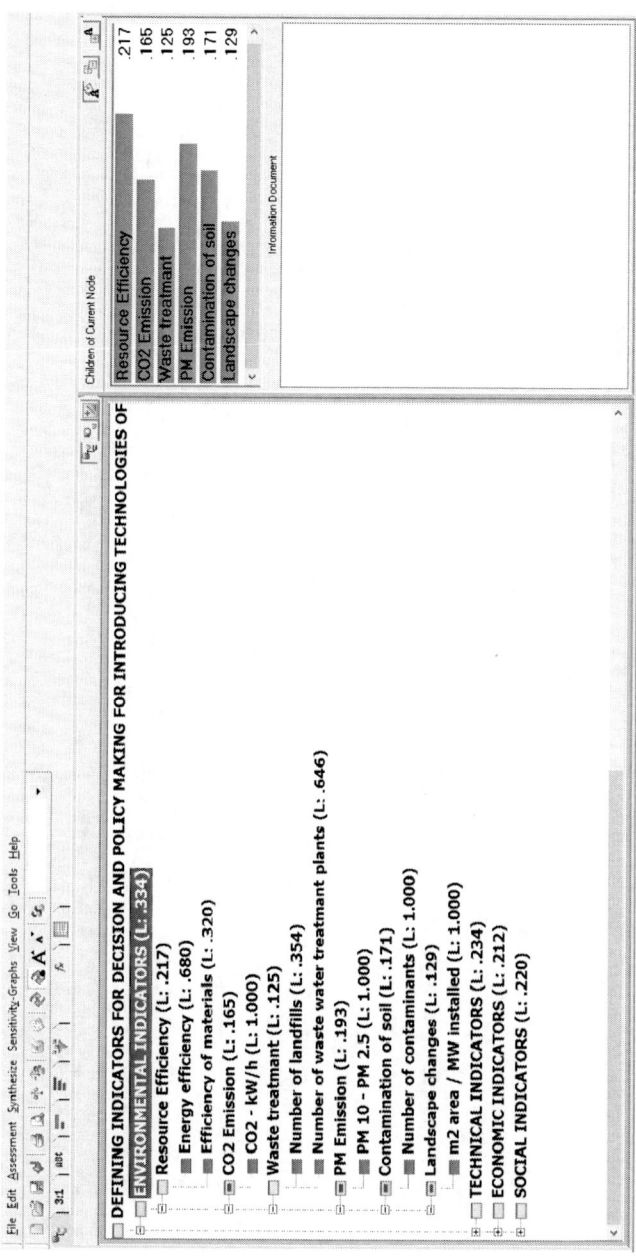

Figure 39. The program interface through which the calculations for Environmental Indicators have been made.

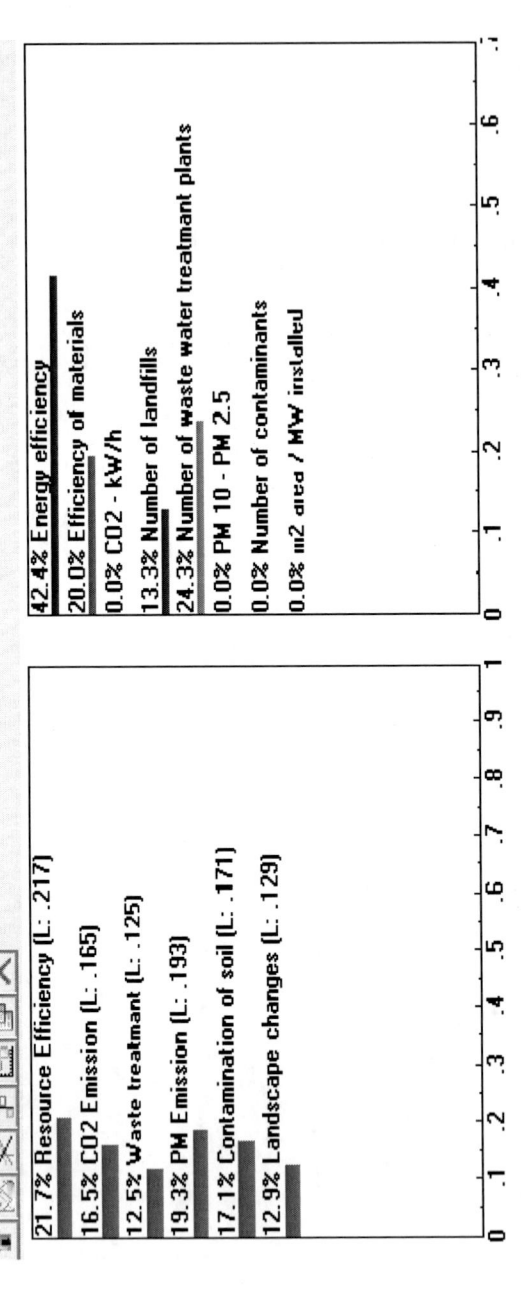

Figure 40. Environmental Indicators, Dynamic Sensitivity.

Technical Indicators

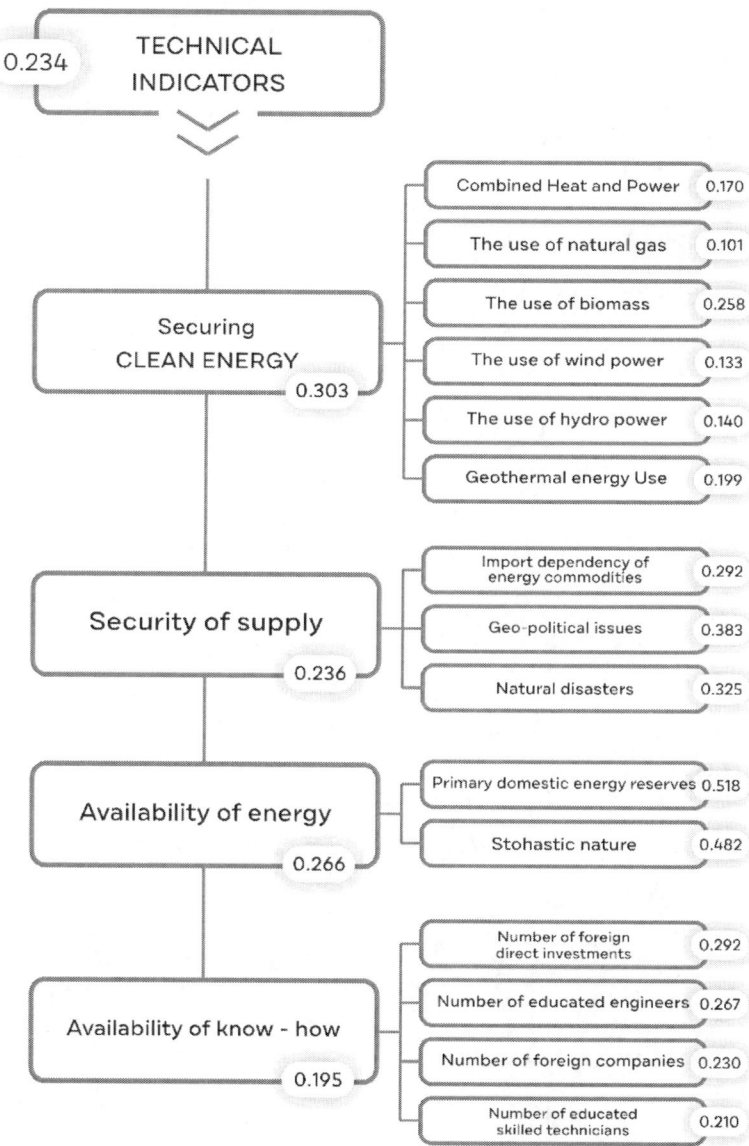

Figure 41. Graphical presentation of results for all hierarchy levels for Technical Indicators.

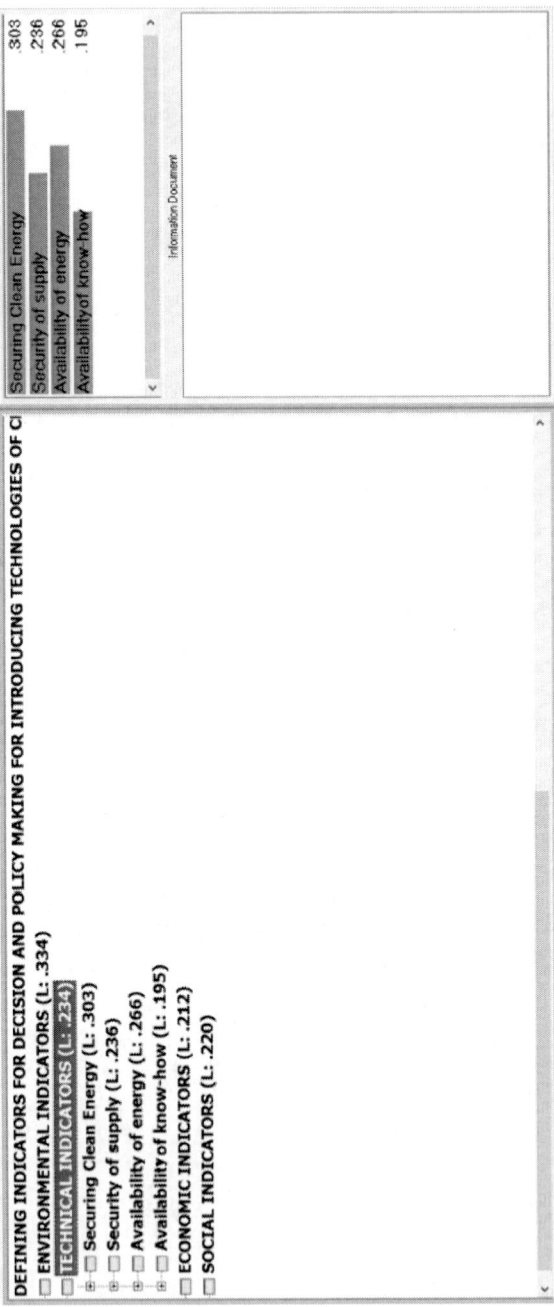

Figure 42. The program interface through which the calculations for Technical Indicators have been made.

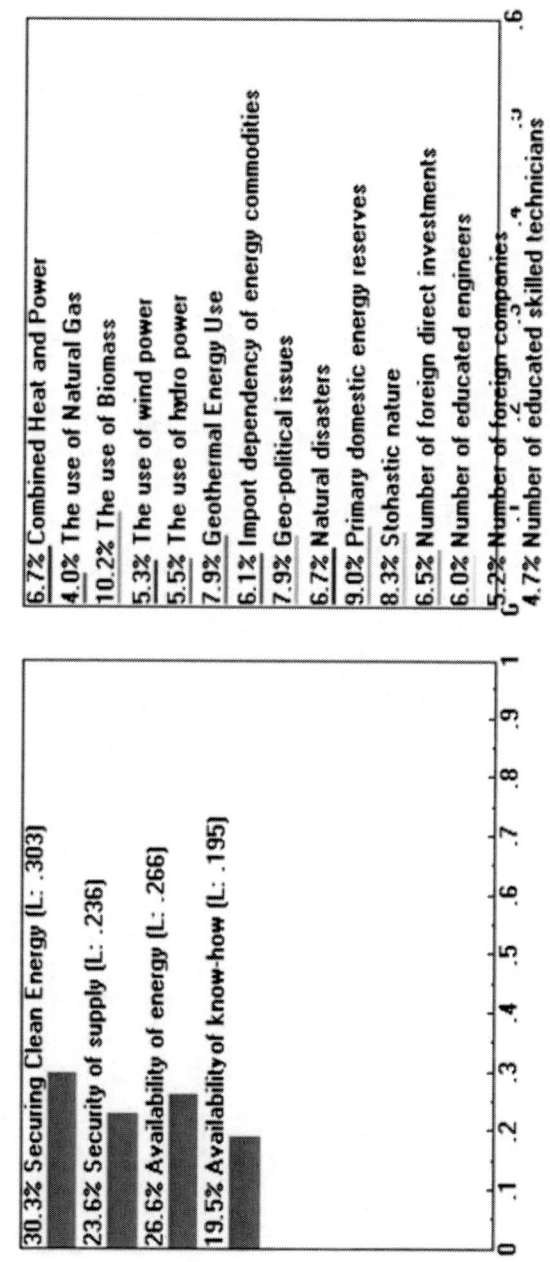

Figure 43. Technical Indicators, Sensitivity Graphs - Dynamic.

Economic Indicators

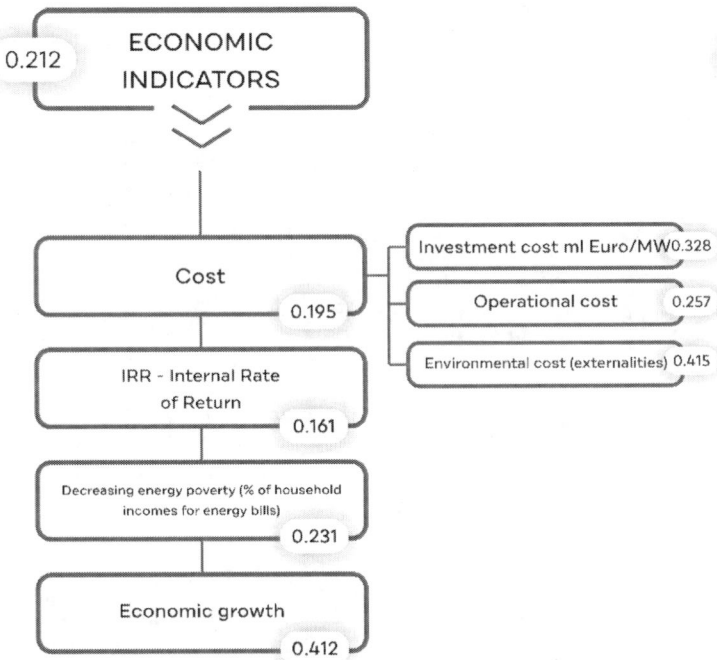

Figure 44. Graphical presentation of results for all hierarchy levels for Economic Indicators.

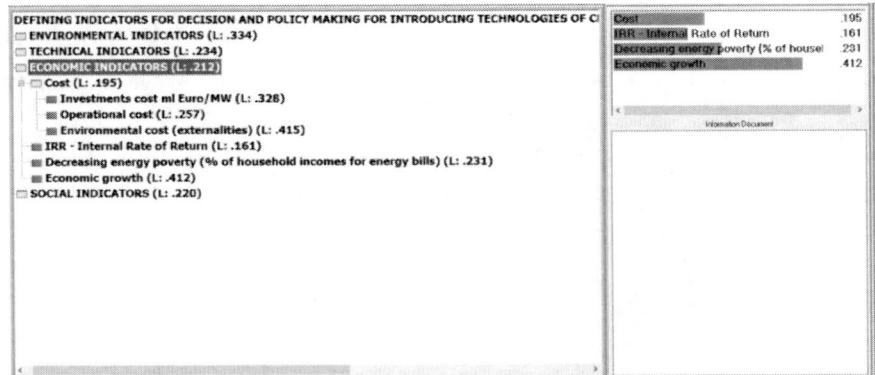

Figure 45. The program interface through which the calculations for Economic Indicators have been made.

Social Indicators

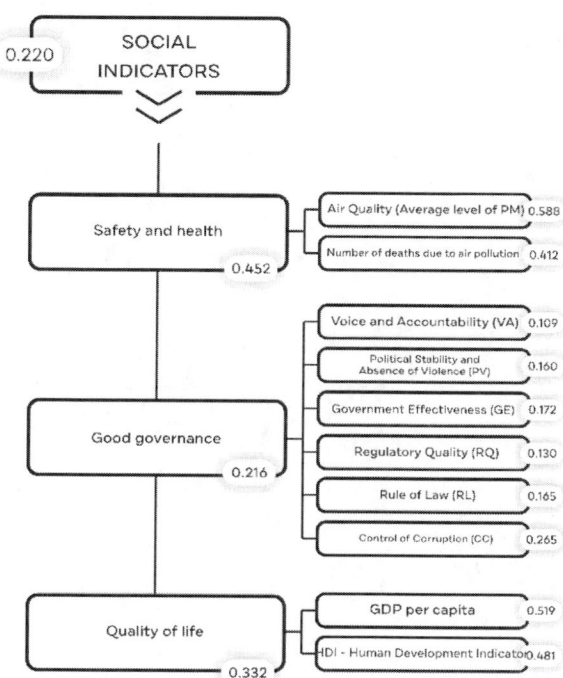

Figure 46. Graphical presentation of results for all hierarchy levels for Social Indicators.

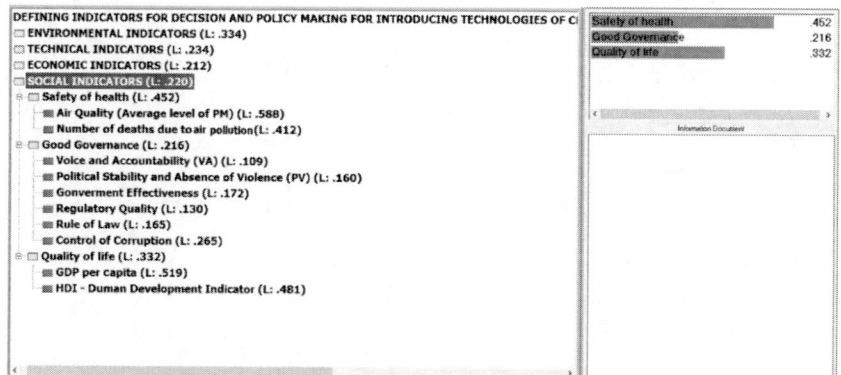

Figure 47. The program interface through which the calculations for Social Indicators have been made.

Table 9. Presentation of alternatives (total 16) through Expert Choice software

Ideal mode Alternative	Total	DIRECT ENVIRONMENTAL INDICATORS CO2 Emission (L: .165)	DIRECT ENVIRONMENTAL INDICATORS Waste treatment (L: .125)	DIRECT ENVIRONMENTAL INDICATORS PM Emission (L: .193)	DIRECT ENVIRONMENTAL INDICATORS Contamination of soil (L: .171)	DIRECT ENVIRONMENTAL INDICATORS Landscape changes (L: .129)	DIRECT TECHNICAL INDICATORS Securing Clean Energy (L: .303)	DIRECT TECHNICAL INDICATORS Security of supply (L: .236)	DIRECT TECHNICAL INDICATORS Availability of energy (L: .266)
☑ RESOURCE	.301	0.35	0.35	0.35	0.35	0.35	0.15	0.35	0.35
☑ CLEANER	.304	0.35	0.35	0.35	0.35	0.35	0.35	0.35	0.15
☑ NEW	.281	0.35	0.35	0.15	0.35	0.35	0.35	0.15	0.15
☑ RENEWABLE	.295	0.35	0.15	0.35	0.35	0.35	0.35	0.15	0.35

Table 10. Presentation of alternatives (total 16) through Expert Choice software

Ideal mode Alternative	Total	DIRECT TECHNICAL INDICATORS Security of supply (L: .236)	DIRECT TECHNICAL INDICATORS Availability of energy (L: .266)	DIRECT TECHNICAL INDICATORS Availabiliti of know-how (L: .195)	DIRECT ECONOMIC INDICATORS Cost (L: .195)	DIRECT ECONOMIC INDICATORS IRR - Internal Rate of Return (L: .161)	DIRECT ECONOMIC INDICATORS Decreasing energy poverty (% of household incomes for energy bills) (L: .231)	DIRECT ECONOMIC INDICATORS Economic growth (L: .412)	DIRECT SOCIAL INDICATORS Safety of health (L: .452)
☑ RESOURCE	.301	0.35	0.35	0.35	0.35	0.35	0.15	0.35	0.35
☑ CLEANER	.304	0.35	0.15	0.35	0.15	0.35	0.35	0.35	0.35
☑ NEW	.281	0.15	0.15	0.35	0.35	0.15	0.35	0.35	0.35
☑ RENEWABLE	.295	0.15	0.35	0.35	0.35	0.35	0.15	0.35	0.35

Table 11. Presentation of alternatives (total 16) through Expert Choice software

Alternative	Total	TECHNICAL INDICATORS Availabiliti of know-how (L: .195)	ECONOMIC INDICATORS Cost (L: .195)	ECONOMIC INDICATORS IRR - Internal Rate of Return (L: .161)	ECONOMIC INDICATORS Decreasing energy poverty (% of household incomes for energy bills) (L: .231)	ECONOMIC INDICATORS Economic growth (L: .412)	SOCIAL INDICATORS Safety of health (L: .452)	SOCIAL INDICATORS Good Governence (L: .216)	SOCIAL INDICATORS Quality of life (L: .332)
RESOURCE	.301	0.35	0.35	0.35	0.15	0.35	0.35	0.35	0.35
CLEANER	.304	0.35	0.15	0.35	0.35	0.35	0.35	0.35	0.35
NEW	.281	0.35	0.35	0.15	0.35	0.35	0.35	0.35	0.35
RENEWABLE	.295	0.35	0.35	0.35	0.15	0.35	0.35	0.35	0.35

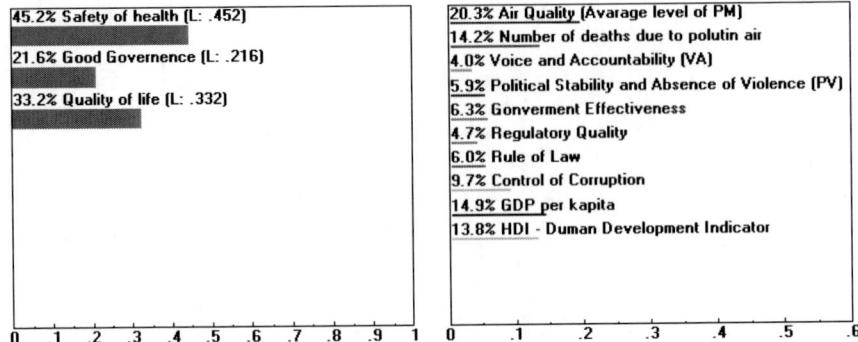

Figure 48. Social Indicators, Sensitivity Graphs – Dynamic.

The multi-criteria method of decision-making, with the use of Expert Choice software, leads to the following results, for all 16 indicators for the two levels of the hierarchy. Our analysis gave the values shown in the tables.

5.2. Multi-Objective Programming Approach Analysis

The application of this method in the research has the aim of providing a more precise quantification and comparison between the most conflicting aspects in electricity generation – the costs (operational costs for production) and the level of CO_2 emissions. In order to make the data comparable, they are expressed in terms of kWh produced electricity. With these assumptions, the formulation of the problem is as follows:

Minimize

$$z(x) = [z_1(x), z_2(x)] \qquad (20)$$

when the functions proposed are:

$$z_1(x) = (28x_1 + 85x_2 + 71x_3 + 77.3x_4 + 136.5x_5)/\sum x_i \quad \text{euro/kWh} \qquad (21)$$

$$z_2(x) = (1015x_1 + 101x_3)/\sum x_i \quad \text{(g/kWh)} \qquad (22)$$

Applied Methodology

The above functions are subject to the following limitations:

$$\sum x_i \geq 6455 \text{ GWh} \tag{23}$$

$x_i \geq 0, \forall i$
$x_2 \leq 260$ GWh
$x_3 \leq 96$ GWh
$x_4 \leq 648$ GWh
$x_5 \leq 232$ GWh

where:
x_1 - energy generation based on coal
x_2 - energy generation based on wind
x_3 - energy generation based on biomass
x_4 - energy generation based on hydropower plants
x_5 - energy generation based on solar energy
$z_1(x)$ - generation cost
$z_2(x)$ - CO_2 emissions

a) Minimizing the cost $z_1(x)$, the following solution is obtained:

Cost, $z_1(x)$: 28.00 euro/kWh and emissions CO_2, $z_2(x)$: 1015 g/kWh with:

x_1=6455 GWh of coal

The interpretation that can be done is to maximize the use of coal, which generates the lowest cost.

b) Minimizing CO_2 emissions $z_2(x)$ results in the following:

Cost $z_1(x)$: 39.78 euro/kWh and CO_2 emissions $z_2(x)$: 822.15 g/kWh with:
x_1 = 5219 GWh of electricity based on coal
x_2 = 260 GWh of wind power
x_3 = 96 GWh of electricity based on biomass
x_4 = 648 GWh of electricity based on hydropower plants
x_5 = 232 GWh of electricity on a solar basis
$z_1(x)$ - generation cost
$z_2(x)$ - CO_2 emissions

In this case, the approach makes the most of all the clean energy sources, leaving the coal to meet the fixed total needs. The obtained values of the optimization with the application of the simplest method are listed in Table 13.

Table 12. Ideal and anti-ideal points used in compromise programming

	Cost, euro/kWh)	Emission of CO2, g/kWh
Cost	28.00	1015.00
Emission of CO_2	39.78	822.15

The main diagonals elements are called ideal locations, giving the solution in which both goals (objectives) reach their optimum value. In reality, the ideal point is unattainable, but it is useful to determine the most appropriate solution in order to homogenize the decision-making units. The most appropriate solution should be chosen from the group of efficient solutions. This group can be approximated by the limitation method.

This method has the purpose of optimizing a goal (objective), including other constraints set as a parametric barrier. For each value of this parameter, a certain point of efficient solutions will be obtained. In this case, in minimizing the cost, for a set of relevant CO_2 emission values, this group is determined by ideal and anti-ideal emission values, for an increase of 10 g/kWh variation. The formulation is as follows:

$$z_1(x) = (28x_1 + 85x_2 + 71x_3 + 77.3x_4 + 136.5x_5)/\sum x_i \quad \text{euro/kWh} \tag{24}$$

$$z_2(x) = (1015x_1 + 101x_3)/\sum x_i \quad \text{g/kWh} \tag{25}$$

subject to the following restrictions:

$$\sum x_i \geq 6455 \text{ GWh}$$
$$x_i \geq 0, \forall i$$
$$x_2 \leq 260 \, G \text{ GWh}$$
$$x_3 \leq 96 \, G \text{ GWh}$$
$$x_4 \leq 648 \text{ GWh}$$
$$z_2(x) = k, \; 822.15 < k < 1015.00 \tag{26}$$

Applied Methodology

The values obtained are shown in Table 13.

Table 13. Efficient set of solutions

Energy Source			COAL	WIND	BIOMASS	HYDRO	SOLAR
Cost of energy production			28.00	85	71	77.3	136.4
Nr	z1(x)	z2(x)	x1	x2	x3	x4	x5
1	39.78	822.15	5219	260	96	648	232
2	38.71	832.15	5283	260	96	648	168
3	37.64	842.15	5346	260	96	648	105
4	36.58	852.15	5410	260	96	648	41
5	35.69	862.15	5473	238	96	648	0
6	35.13	872.15	5537	174	96	648	0
7	34.56	882.15	5601	110	96	648	0
8	34.00	892.15	5664	47	96	648	0
9	33.46	902.15	5728	0	96	631	0
10	32.97	912.15	5791	0	96	568	0
11	32.49	922.15	5855	0	96	504	0
12	32.00	932.15	5919	0	96	440	0
13	31.52	942.15	5982	0	96	377	0
14	31.03	952.15	6046	0	96	313	0
15	30.55	962.15	6109	0	96	250	0
16	30.06	972.15	6173	0	96	186	0
17	29.57	982.15	6237	0	96	122	0
18	29.09	992.15	6300	0	96	59	
19	28.60	1002.15	6364	0	91	0	
20	28.13	1012.15	6435	0	20	0	
21	28.00	1015.00	6455	0	0	0	

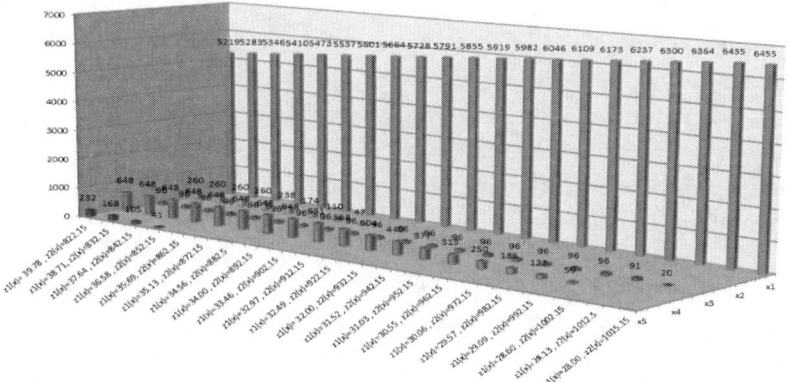

Figure 49. Graphical presentation of efficient set of solutions.

In this table, it can be seen that various items participate in the generation of electricity for the options being dealt with. It can be noticed how many different options enter the optimal solution according to the changing of the cost of CO_2 generation and emission. Thus, for example, wind energy is the only solution when CO_2 emissions are limited to 902.15 g/kWh, bringing the cost of production to 33.46 euro/kWh. However, if the cost of generation is limited, for example, 30.55 euro/kWh, there is only room for the mixed optimum generation of coal, biomass and hydropower.

With the values obtained, a chart is developed, in which the ideal point is also noted.

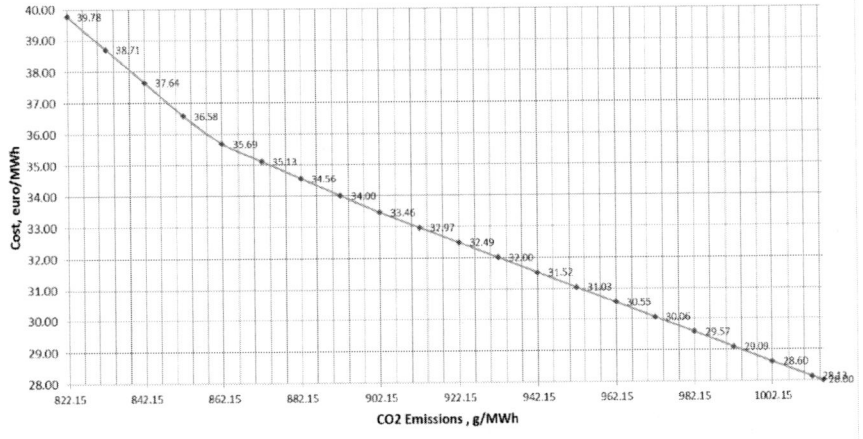

Figure 50. Compromise group for equal weight criteria.

All the solutions represented on the curve graph are viable and efficient solutions. But to solve the problem it is necessary to choose just one of the items. This requires the input preferences of the decision-making center, for these will determine the solution, in view of the importance of achieving each objective (purpose), in our case the value of the minimum generation cost or the minimum emission of CO_2, where the optimal point will move along to the right or left of the curve.

Therefore, the displayed curve assists in decision-making because it shows all of the efficient optimization solutions to the problem as a means to find the solution that best suits the interests of the decision-makers. Thus, from the group of solutions, it is possible to choose an optimal solution, by means of compromise programming, as described below.

Compromise Programming

Compromise programming was used to select the optimal element from a range of efficient solutions as proposed by Zeleny. This rule is called Zeleny's axiom and is expressed as follows: "Taking into consideration solutions in target space z1 and z2, the preferred solution will be the one closest to the ideal point" (Cochrane and Zeleny 1973) [87].

If we assume that the decision-making center behaves in a rational way, it will select the effective point or zone of effective point groups that is closest to the ideal point.

Compromise programming begins by setting the ideal point, the coordinates of which are given by the optimum values of the various objectives of the decision-maker. The ideal point is usually inadequate. If possible then there is no conflict between the objectives. When the ideal point is inadequate, the optimal elements or compromise solutions are provided by an efficient solution that is closest to the ideal point.

The ideal alternative coordinates are given by optimal values usually when targets (goals) are measured in different units, so that the degree of proximity does not make sense or have any meaning because of the lack of dimensional homogeneity. Therefore, it will be necessary to continue with the normalization of objectives (goals). Thus, the degree of proximity as the relative deviation between the objective j and its ideal value is determined by [88]:

$$d_j = \frac{\left[z_j^* - z_j(x)\right]}{\left[z_j^* - z_{*j}\right]} \tag{27}$$

where d_j represents the degree of proximity of the normalized objective j and z^*j is the anti-ideal of this objective, the worst possible value for objective j in an effective set (efficiency). The normalized degree of proximity is limited between 0 and 1. Thus, when a target reaches its ideal value, its proximity is zero; on the contrary, this scale becomes equal to one when the objective in question reaches a value equal to the anti-ideal. If we now represent the W_j preferences that the decision-making center relates to the discrepancy between achieving objective j and its ideal goal, compromise programming is consistent in seeking more efficient solutions closer to the ideal. So this programming is based on whether the optimal solution is to find the point closest to the ideal. This proximity is measured by the mathematical concept of distance.

There are many types of distance, apart from the Euclidean [89], which is the best known, and the question is which to use. In fact, the process can be simplified, as it is shown that the range of distance solutions from the ideal point is the one that is minimal, namely the so-called Manhattan *L1* [90], and Chebyshev *L∞* [91] distances.

These points are unchanged as they depend on the given weighting of each known target and must reflect the preferences of the decision-making center. These weights are placed in the expression of the distance, so that the result is as follows:

$$L_1 = W_1 \frac{z_1(x) - z_1^*}{z_{*1} - z_1^*} + W_2 \frac{z_2(x) - z_2^*}{z_{*2} - z_2^*} \tag{28}$$

$$L_\infty = \max[W_1 z_1(x) - z_1^*, W_2 z_2(x) - z_2^*] \tag{29}$$

where:

z_j^* : the ideal value

z_{*j} : the anti-ideal value

Wj : the weight of each objective

For example, we have calculated compromise groups for different weights. Here we consider sharing the same weight for both objectives, one twice as important as the other, and also one four times more important than the other. The groups that result in compromise are reflected in Table 13.

The part of each type of energy in the optimal solution can be seen, for each case in Table 14, by introducing the cost of generation and the corresponding CO_2 emissions. As an illustration, here in detail a compromise is made for the case of the same importance for two objectives that are considered certain.

Table 13. Compromise solutions L1 and L∞

L1	35.46 euro/MWh	866.10 g/MWh
L∞	36.00 euro/MWh	856.59 g/MWh

Table 14. Compromise solutions for energy sources L1 and L∞

Energy Sources	Distance Li	Distance L∞
Coal	5499	5483
Wind	212	228
Biomass	96	96
Hydraulics	648	648
Solar	0	0

As it can be observed, wind energy appears in this group of compromises, though with a slightly lower turnout. However, solar energy does not appear, because the cost of generation is too high, as shown in Table 14.

Discussion of Analysis and Conclusion

All the herein performed analysis in terms of indicators is based on using the AHP method, through which reliable and practically appropriate results are obtained, particularly when many criteria are involved in decision-making.

Using the AHP method, of the four alternatives, besides Cleaner Production, Resource Efficiency, and New Technologies, the Renewable Energy alternative, which through the analysis carried out proved to be the most expensive method to use, had more advantages in terms of environmental protection. This result, of course, can be read from the hierarchy diagram of the problem.

The approach of applying linear programming in decision-modeling, in this case concerning energy generation, relies on some basic assumptions about modeling the situation, and the decision-maker in theory aims to choose a well-defined target. In reality, currently the decision-maker usually requires an effective compromise between several objectives, many of which may be in conflict, or tries to achieve satisfactory levels of the goals [92, 93].

Models that incorporate multiple goals (purposes) are useful in researching and expanding power generation systems. As we have presented in the previous chapters, currently different authors prefer different methods of MCDM (Multi-Criteria Decision-Making) such as the AHP method, TOPSIS, Pareto, ELECTRE, PROMETHEE, Compromises etc. The compromise programming technique [94] can be used in studies to find compromise solutions between the various contradictory objectives in the electricity generation system, such as minimizing cost and CO_2 emissions [95, 96, 97]. In this case, electricity generation in the Republic of Kosovo has been addressed.

The generation of electricity has environmental impacts, however, which are not taken into account when allocating resources efficiently. So change to the environment continues to happen, despite growing public awareness, and market forces have no means to prevent it. In order to avoid these weaknesses, multi-criteria programming is used. This method has the advantage that it is not necessary to determine a monetary value for environmental impacts, as it allows the analysis to include heterogeneous units.

The case of simplified electricity planning has been analyzed with some assumptions made about energy planning in order to reduce the operational complexity. This simplification is merely meant to show how the results of the traditional electrical planning approach can be changed if other criteria are presented in addressing the case of electricity generation in the Republic of Kosovo (RK). The introduction of environmental costs is done through a related variable, in this case CO_2 emissions, of environmental sustainability.

In the second part of the analysis/research, the four alternatives are defined. Due to the importance of the indicators addressed in the second level of the hierarchy, it is judged that all these indicators (16 in total) should be compared to all alternatives. Such logic has been given as a special part of the questionnaire built for the problem, in which the experts involved in the research subsequently gave their professional opinions.

The results obtained provide a realistic picture of the directions to be taken in order to improve the situation in the analyzed areas. As shown by the Environmental Indicators, sustainable development to a large extent depends on the Resource Efficiency indicator (0.217), as a step to be taken by decision-makers and policy-makers in the institutions of the Republic of Kosovo. This will help the industrial sector in general, with special emphasis on small and medium enterprises, taking into account that the biggest problem they face is related to affordable energy, whereas Resource Efficiency as a practice has no application at all.

From the perspective of the professionals who gave their assessment of the hierarchy of the problem, it can be noticed that Resource Efficiency, as an alternative to sustainable development, is an important part of decision-making and policy-making, being ranked as the second alternative (0.301).

PM Emission (0.193), Contamination of soil (0.171), CO_2 Emission (0.165), Landscape Changes (0.129) and Waste Treatmant (0.125) comprise the rest of the Environmental Indicators. The great need for politicians and decision-makers to intervene to correct the indicators in the near future will certainly help the Republic of Kosovo to regulate the environmental parameters in the move to sustainable development and improvement of the environment, turning these into "normal" and acceptable parameters. Waste treatment, emission of pollutants, land contamination and landscape shift are related to the last indicator of our Social Indicator analysis, which is closely related to the "performance" of environmental parameters. The change of each indicator, meaning the advances made in each field, even at a minimal level, will have an impact on the whole hierarchy of the problem, making the model applicable and usable even in the search for practical results.

Based on the data introduced in the model, a dominant indicator from the third level at the Environmental Indicators proved to be Energy Efficiency (0.680). Certainly, this result was as expected, given the high and unbearable cost of electricity "for industry" as well as the lack of a thermal energy network. So, the indicator that needs to be addressed in a sensitive way has to do with the increase of Energy Efficiency as an acceptable way to provide an affordable energy supply. Energy is a fundamental part of the operation of SMEs, and as such is a fundamental part also of the cleaner production process.

Securing Clean Energy (0.303) is the most important indicator in the whole second level of the problem hierarchy of technical factors. Statistically, the basic problem in the functioning of industry and enterprises of all levels is that of power supply. In terms of planning to improve the parameters for Securing Clean Energy, Kosovo has drafted/approved strategic documents, but based on this research the results show that this work is dealing with the most important indicators through which a sustainable stability and development will be achieved.

Economic indicators are of great importance not only in this analysis as a theoretical treatment but also at the practical level. The indicators, whose weight was tested, yielded the obviously expected results, expressing their dominance through Economic Growth (0.412). Economic Growth has a fair relationship with the performance of industry and enterprises which operate at the country level. Improving the performance of this industry and these companies will achieve a "reflection" of the economic indicators that guarantee sustainable development. Attempts to improve the economic indicators should be made in each case by addressing the four alternatives in a serious, professional and competitive way through the process of decision-making and policy-making.

The opinion of professionals regarding social indicators was required for 13 indicators at the second and third levels of the hierarchy. The Safety and Health indicator (0.452) naturally proves to be the dominant indicator in the second level of the hierarchy, showing security and health as the most important issues in the domain of social indicators. Air Quality (0.588) should be treated with great care and priority as one of the indicators that can affect the improvement of Safety and Health by reducing the number of deaths due to air pollution (0.412). Numerous studies indicate that the number of premature deaths in the region caused in particular by coal power plants is extremely serious, as well as the incidence of respiratory and other deseases related with the polluted air. Within the second level of the hierarchy, Quality of Life (0.332) is ranked after Safety and Health, followed by Good Governance. The quality of life is a complex indicator which embeds the GDP value, as well as the Human Development Index (HDI).

The good governance of each country plays a very important/fundamental role in sustainable development and not only in the wellbeing of citizens. Good governance in this case has been analyzed through six "standard" indicators on which it depends. Control of Corruption (0.265) is the main indicator through which good governance is achieved and is thus pursued in sustainable development. Government Indicators (0.172), Rule of Law (0.165), Political Stability and Absence of Violence (0.160) and Voice and Accountability (0.109) are the other five indicators which, with different weighting factors, play an important role in improving the governance.

The Recommended Application of the AHP Method

Throughout the analysis and explanations given in this book, the use of the MCDM – AHP methodology has been introduced to define the indicators that contribute to decision-making and policy-making for sustainable development in the field of energy planning, taking Kosovo as a case study, which can then be compared to the Western Balkans states as "developing industries."

Indicators have been identified that have an impact on sustainable development in cleaner production, as a field that the Republic of Kosovo needs to address through a clear agenda for the steps to be taken in order to move forward. Enterprises in general and SMEs in particular face the major challenge of energy security, large amounts of waste (which in the further stages should be treated as a reusable asset/energy production) and high emission levels, resulting in environmental pollution.

For the method used herein, the hierarchy of the defined problem was originally set up based on four categories (Environmental, Technical, Economic and Social) under which all the indicators are grouped. The purpose of each category is to identify as many relevant indicators as possible for the problem being addressed. The weight factors for each indicator are calculated through the AHP method, one of the most commonly used methods in decision-making and policy-making, especially when it comes to situations requiring decisions to be taken in a short time.

From the analysis conducted, the contribution of each analyzed indicator can be seen: environmental, technical, economic and social, in relation with other indicators. The environmental factor has the greatest contribution compared to the technical, economic and social factors. Technical, economic and social factors show no major differences in their ranking. Overall, on the basis of the values achieved, we see an advantage of social factors in relation to economic ones. The advantage of social factors is a result of the "safety and health" indicator on social factors, which made the difference in this regard. This finding is not a pure accident, since both groups of indicators (social and environmental) have great importance in terms of the analysis of so-called "external costs" or externalities – implicit costs due to the negative impacts on the environment and society. These costs are a burden for the state, and its citizens, through increased costs in terms of public health issues caused by environmental pollution. Thus, the results of the analysis show the importance of each factor at all levels of the hierarchy in the model. Through these results it can be seen that the most important indicators are Resource Efficiency, Energy Efficiency, Securing Clean Energy, Safety and Health, Economic Growth, Number of Foreign Investments, Good Governance, and Control of Corruption, which makes clear the unsustainable situation and the need for Kosovo in particular and the countries of the region in general to pursue sustainable development, right decision-making and well-planned policies. The indicators addressed in the environmental field provide a clear and realistic picture of the steps tht need to be taken to further improve the environmental situation, which is also crowned by the Cleaner Production alternative as a route to follow in order to ensure sustainable development.

The need for affordable and clean energy as one of the UN projections (out of 17) has been manifested as a result of the technical indicator analysis, where Securing Clean Energy, Availability of Energy, Security of Supply and Availability of Know-how are in the forefront. Through the interconnection with "Cleaner Production", which is found to be the dominant alternative (out of the four), the need can be seen to implement this in all types of industries/enterprises, and in small and medium enterprises in particular.

Safety and Health, Good Governance and Quality of Life, as three of the analyzed indicators in social factors, indicate the need for the country to seriously address these issues. Naturally, the Safety and Health indicator is dominant in relation to the other two indicators. The results of the analysis for the Good Governnance and Quality of Life indicators highlight the tremendous weakness in the country's governance and the need to improve the situation in the field of Control of Corruption and Governance Effectiveness.

Other indicators through which the power supply opportunities of the Republic of Kosovo have been presented give results indicating the direction in which it should be heading. Securing Clean Energy is enabled by advancing/stimulating energy production based on Renewable Energy. This and New Technologies are defined as two of the four sustainable development alternatives but differ a little from Cleaner Production and Resource Efficiency as alternatives, and therefore lagging slightly behind.

Since energy is the cornerstone of sustainable development today, it has a dominant role in policy-making and decision-making. This is why the Geopolitical Issues Index (0.383) is dominant among the Security of Supply indicators, followed by Natural Disasters. Political developments at the national, regional and world/international level have a great impact on small markets such as Kosovo and the Balkans, which cannot dictate major changes in energy. Good cooperation and correct/stable policy-making would guarantee a sustainable energy supply as a prerequisite for sustainable development.

In order to perform more accurate analyses of the interrelations between economic direct (explicit) costs and indirect (external) costs (social, environmental), multi-objective programming, i.e., compromise programming, is applied, by using two opposite criteria in the goal function – minimum investment costs and minimum CO_2 emissions. This approach provides a ranking of the several different energy generation sources in terms of overall production costs.

To achieve accuracy and transparency in the decision-making process for sustainable development of the country by advancing the idea of using Cleaner Production and Resource Efficiency, especially when it comes to applying high standards of use of Energy Efficiency and New Technologies, the professional opinions of experts of the respective fields must be respected.

Thus, the applicable method is the multi-criteria approach, which prevents subjectivity among all actors. So, it is the only verified way that justifies the use of the optimal scenario for sustainable development in the country and in the region. This case study offers interesting trajectories of the problem, through which a number of indicators related to sustainable development are identified, also offering certain alternatives to determine the path that enables the choice of the optimal alternative.

The results obtained from the analysis made in the area of decision-making and policy-making, which address a large number of indicators determined through the model, may also be influenced by the consistency of the experts participating in the evaluation of the indicators. From the assessment of some opinions, a slight lack of consistency in the evaluation of indicators is noticed, but this does not represent any major difference that would affect the final outcome.

This analysis has contributed to the opening of many new questions and also new requests for inquiries. Our analysis case, which is the Republic of Kosovo in particular and the region in general, needs to address these issues in the future in order to define more indicators that have an impact on the sustainable development of this whole region.

The next phase of research necessarily has to focus on the analysis by combining different scenarios of problem hierarchies, in some cases reducing the number of indicators in the hope that the number of non-consistent responses will be smaller, and in other cases increasing their number in order to deepen the analysis, taking into account that we are dealing with the multi-criteria method of problem-solving.

Further research is needed in the field of determining the dependencies between different indicators. The concept presented in this analysis can be a solid foundation for this, given that Kosovo so far has little to say about such studies in the field of decision-making and policy-making for Cleaner Production and Resource Efficiency.

Using the AHP method, with the idea of setting the indicators, their weight, defining the hierarchy of the problem up to ranking and offering alternatives naturally, does not imply that it is a method through which we can definitively solve the real problems that mankind faces nowadays. But, by this method it is feasible to define the future goals use and application of other methods in the field of decision-making and multi-criteria policy for the Republic of Kosovo in projects that enable sustainable development.

However, the applicability of this model (with certain minimal modifications) in terms of other countries in the region, as well as over similar problem definitions, enables the preconditions for its wider acceptance. As per the term "wider", this comprises the institutional inclusion of proven scientific methods and techniques in decision-making, as well as mapping of the already created models and studies from one country to another in the Western Balkans region. Thus, it will contribute towards better regional cooperation in the treatment of similar problems (in particular in the area of sustainable development), and it will foster the preparations of the countries for the pre-accession negotiations for EU membership, which is one of the main common goals of the entire region.

References

[1] United Nations Conference on Environment & Development Rio de Janeiro, Brazil, 3–14 June 1992. https://sustainabledevelopment.un.org/content/documents/Agenda21.pdf (accessed 23.02.2018).

[2] http://www.un.org/sustainabledevelopment/sustainable-development-goals/ (accessed 12.02.2018).

[3] Strange, T. and Bayley, A. (2008). Sustainable development: linking economy, society, environment. OECD Insights.

[4] International Institute for Sustainable Development. http://www.iisd.org/topic/sustainable-development (accessed 16.02.2018).

[5] United Nations. Our Common Future, 1987. Report on the World Commission on Environment and Development. http://www.un-documents.net/our-common-future.pdf (accessed 22.03.2018).

[6] Hansmann, R., Mieg, H. A. and Frischknecht, P. (2012). Principal Sustainability components: empirical analysis of synergies between the three pillars of sustainability. *International Journal of Sustainable Development & World Ecology* 19 (5): 451-459.

[7] Coase, R. H. (1960). The Problem of Social Cost, Journal of Law and Economics 3: 1-44.

[8] Partnerships for Sustainable Development Goals. https://sustainabledevelopment.un.org/content/documents/211617%20Goals%2017%20Partnerships.pdf (accessed 03.02.2018).

[9] United Nations. Transforming Our World: The 2030 Agenda for sustainable Development. https://sustainabledevelopment.un.org/content/documents/21252030%20Agenda%20for%20Sustainable%20Development%20web.pdf (accessed 03.02.2018).

[10] https://www.researchgate.net/post/What_is_the_best_definition_for_sustainable_development (accessed 10.02.2018).

[11] Swiss Federal Office for Spatial Development. 1987: Brundtland Report. https://www.are.admin.ch/are/en/home/sustainable-development/international-cooperation/2030agenda/un-_-milestones-in-sustainable-development/1987--brundtland-report.html (accessed 30.01.2018).

[12] United Nations General Assembly (2015). Resolution adopted by the General Assembly on 25 September 2015. Seventieth session, Agenda items 15 and 116.

[13] Environmental policy document catalogue (2015) – Rozolution adopted. www.eea.europa.eu
Resolution adopted by the General Assembly on 25 September 2015. Transforming our world: the 2030 Agenda for Sustainable Development

[14] Suhrke, A. (1994). Environmental Degradation and population flows. *Journal of International Affairs* 47 (2): 473-496.

References

[15] Tanazian, A., Chousa, J. P. and Vadlamannati, K. C. (2009). Does higher economic and financial development lead to environmental degradation: evidence from BRIC countries. *Energy Policy* 37 (1): 246-253.

[16] Tyagy, S., Gaug, N. and Paudel, R. (2014). Environmental Degradation: causes and consequences. *European Researcher* 81 (8-2).

[17] Rosu, A., Constantin, D. E. and Georgesku, L. (2016). Air Pollution Level in *Europe Caused by Energy Consumption and Transportation*. J Environ Prot Ecol 17 (1): 1.

[18] Organisation for Economic Co-operation and Development (OECD). (2008). Key Environmental Indicators. OECD Environment Directorate Paris, France.

[19] Ministry of Environment and Spatial Planning & Kosovo Environmental Protection Agency. (2017). *Annual Report State of the Environment in Kosovo*, Prishtine.

[20] UN Sustainable Development Goal 7: http://www.undp.org/content/undp/en/home/sustainable-development-goals/goal-7-affordable-and-clean-energy.html (accessed 30.01.2018).

[21] Organisation for Economic Co-operation and Development/International Energy Agency United Nations Development Programme (2008). *Energy in the Western Balkans: The Path to Reform and Reconstruction*. International Energy Agency.

[22] Golušin, M., Munitlak Ivanović, O. and Redžepagić, S. (2013). Transition from traditional to sustainable energy development in the region of Western Balkans – Current level and requirements. *Applied Energy* 101: 182-191.

[23] Cherp, A., Jewell, J. and Goldthau, A. (2011). Governing Global Energy: Systems, transitions, Complexity. *Global Policy* 2(1):75-88.

[24] Estrin, S. and Uvalić, M. (2013). Foreign Direct Investment into transition economies: Are the Balkans different? LEQS Paper No. 64. London School of Economics and Political Science.

[25] Spreng, D. and Wils, A. (1996). Indicators of Sustainability; Indicators in Various Scientific Disciplines. (http://e-collection.library.ethz.ch/eserv/eth:24979/eth-24979-01.pdf). (accessed 22.03.2018).

[26] Berthomieu, C., Cingolani, M. and Ri, A. (June 2016). Investment for Growth and Development in the Western Balkans. STAREBEI Research Project EIB Institute University of Nice – Sophia Antipolis (France).

[27] World Bank (2011). World Development Indicators 2011. World Bank. http://siteresources.worldbank.org/DATASTATISTICS/Resources/wdi_ebook.pdf (accessed 22.03.2018).

[28] European Stability Initiative, 2002. EU Pillar – PISG – Energy Office. Energy strategy and Policy of Kosovo. (White paper). (http://www.esiweb.org/index.php?lang=en&id=27&cat_ID=10). (accessed 22.03.2018).

[29] Black & Veatch Corporation (2012). Cost report: Cost and performance data for power generation technologies. Prepared for the National Renewable Energy Laboratory. Black & Veatch Corporation (Ed.), Kansas, USA.

[30] WEC (2013). World Energy Perspective: Cost of Energy Technologies (Project partner: Bloomberg New Energy Finance). *World Energy Council* (WEC), London, UK, Bloomberg New Energy Finance, New York, USA.

[31] HEAL (2016). The unpaid Health bill – how Coal power plants in the Western Balkans make us sick. Jensen, K. G. (Ed.), Health and Environment Alliance

(HEAL), Brussels, Belgium. (http://env-health.org/IMG/pdf/factsheet_eu_and_western_balkan_en_lr.pdf). (accessed 23.03.2018).
[32] Buzar, S. (2007). *Energy poverty in Eastern Europe*; Hidden Geographies of Deprivation.
[33] Grausz, S. (2011). *The social costs of coal: Implications for the World Bank*. (www.climateadvisers.com), Washington DC, USA.
[34] Goode, N. (Ed.) (2015). More for less? The case of social value for public service delivery. Aventia Consulting Limited with the assistance and support of the Social Value Portal, Edinburgh, UK.
[35] Thomas, M. A. (2009). What Do the Worldwide Governance Indicators Measure?, *European Journal of Development Research*, 22, pp. 31–54, doi:10.1057/ejdr.2009.32, http://iis-db.stanford.edu/docs/623/Thomas_Gov_Indicators.pdf
[36] Raju, K. S. and Kumar, D. N. (2010). Multicriterion analysis in Engineering and Management. PHI Learning.
[37] м-р Даниела Младеновска, дипл. маш. инж.: Определување на индикатори за донесување одлуки и креирање политики во синџирите за снабдување во услови на пазарна економија во Република Македонија Докторска Дисертација. Машински факултет Скопје, Универзитет Св. Кирил и Методиј [Daniela Mladenovska, M.Sc., B.Sc. mash. eng: *Determining indicators for decision making and policy making in supply chains in a market economy in the Republic of Macedonia*. Doctoral Dissertation. Faculty of Mechanical Engineering Skopje, Ss. Cyril and Methodius.].
[38] Miller, G. (1956). The magical number seven, plus or minus two: some limits on our capacity for processing information. *The Psychological Review* 63: 81-97.
[39] Saaty T. L. (1994). How to make a decision: The Analytic Hierarchy Process. *Interfaces* 24: 19-43.
[40] Saaty, T. L. (1986). Axiomatic Foundation of the Analytic Hierarchy Process. *Management Science* 32 (7): 841-855.
[41] Saaty, T. L. (1990). How to make a decision: The Analytic Hierarchy process. *Eur J Oper Res* 48: 9-26.
[42] Sutter, C. (2003). Sustainability Check-up for CDM projects: How to assess the sustainability of international projects under the Kyoto Protocol; Swiss Federal Institute of Technology Zurich 2003; Part II: Theory.
[43] Zardari, N. H., Ahmad, K., Shirazi, S. M. and Yusop, Z. B. (2015). XI Weighting Methods and their effects on Multi Criteria Decision Making outcomes in Water Resources Management.
[44] Rogers, M., Bruen, M. and Maystre, L. Y. (2000). ELECTRE and Decision Support Methods and Applications in Engineering and Infrastructure Investment. Kluwer Academic Publishers, The Netherlands.
[45] Bargueño, D. R., Pamplona Salomon, V. A., Silva Marins, F. A., Palominos, P. and Marrone, L. A. (2021). State of the Art Review on the Analytic Hierarchy Process and Urban Mobility. *Mathematics* 9: 3179. MDPI Journal/Mathematics.
[46] Stojanivic. M. Multi-citeria decision-making for selection of renewable energy system (2013) – *Safety Engineering*. Vol 3, No 3 (2013) 115-120.

References

[47] Luc, D. T. (2013). *Multiobjective Linear Programming*. Springer.
[48] Santos Silva, C. A. (2018). Multiobjective Optimisation. MIT Portugal. https://fenix.tecnico.ulisboa.pt/downloadFile/3779575589383/Class8 (accessed 19.02.2018).
[49] Pandian, P. and Jayalakshmi, M. (2013). Determining Efficient Solutions to Multiple Objective Linear Programming Problems. *Applied Mathematical Sciences* 7 (26): 1275-1282.
[50] Reynolds, J. H. and Ford, E. D. (1999). Multi-Criteria Assessment of Ecological Process Models. *Ecology* 80 (2): 538–553.
[51] Klaassen, G. and Riahi, K. (2007). Internalizing externalities of electricity generation: An analysis with MESSAGE-MACRO. *Energy Policy* 35 (2): 815-827.
[52] Burtraw, D. and Krupnick, A. (2012).The True Costs of Electric Power. *Summary for Policy Makers*. REN21.
[53] PHPSimplex. Optimizing Resources with Linear Programming. http://www.phpsimplex.com/en/. (accessed 21.02.2018).
[54] Solving LPs: The Simplex Algorithm of George Dantzig. https://sites.math.washington.edu/~burke/crs/407/notes/section2.pdf (accessed 22.02.2018).
[55] Afgan, N. H., Al Gobaisi, D., Carvalho, M. G. and Cumo, M. (1998). Sustainable energy development. *Renew. Sustain. Energy Rev.* 2: 235-286.
[56] Wang, T. C., Liang, L. J. and Ho, C. Y. (2006). Multicriteria decision analysis by using fuzzy VIKOR. *Proceedings of International Conference on Service Systems and Service Management*, 2: 901-906.
[57] Mladenovska, D. and Kochov, A. (2017). Assessment of Alternatives for Natural Gas supply in Macedonia versus Technical Indicators. Chapter 12 in: *Advances in Production and Industrial Engineering* (Eds. Čuš, F. and Gečevska, V.), University of Maribor Press. (http://press.uni.si).
[58] Afgan, N. H., Pilavachi, P. A. and Carvalho, M. G. (2007). Multicriteria evaluation of natural gas resource. *Energy Policy* 35: 704-713.
[59] Mladenovska, D. and Lazarevska, A. M. (2015). Decision making concept for creating policies for natural gas supply chain in Macedonia; SDEWES Dubrovnik, 27 September – 03 October 2015.
[60] Boardman, B. (1991). *Fuel poverty: From Cold Homes to Affordable Warmth*. London.
[61] Bouzarovski, S. (2010). Energy poverty in transition: Macedonia and the Czech Republic in comparative perspective, *Political Thought, Year 8*, No. 29, (Skopje, 2010).
[62] Hossain, T. (2015). Application of Resource Efficient and Cleaner Production (RECP) in the Energy Intensive Industry to Promote Low Carbon Industrial Development in Bangladesh. Low Carbon Economy 6: 73-85.
[63] Ackermann, R. et al. (1998). Pollution prevention and abatement handbook 1998: toward cleaner production.
[64] Joint UNIDO-UNEP Programme on Resource Efficient and Cleaner Production in Developing and Transition Countries. http://www.unep.fr/scp/cp/pdf/RECP%20Programme%20Flyer%20April%202010.pdf (accessed 04.02.2018).

References

[65] UNEP/UNIDO (2016). (http://mineacom.gov.rw/fileadmin/templates/Documents/Announcement/Submitted_Final_Report_RECP_Guidelines_v11.pdf) (accessed 06.02.2018)

[66] UNEP (2016). http://www.recpnet.org/wp-content/uploads/2016/10/RECPnet_Charter.pdf (accessed 05.02.2018).

[67] Ellen MacArthur Foundation: Intelligent Assets - Unlocking the circular economy.

[68] Fresner, J. (2017). *Resource Efficient and Cleaner Production Investment Guidelines for New Industries. Final report* part 1. http://mineacom.gov.rw/fileadmin/templates/Documents/Announcement/Submitted_Final_Report_RECP_Guidelines_v11.pdf (accessed 07.02.2018).

[69] Nilsson, L., Persson, P. O., Rydén, L., Darozhka, S. and Zaliauskiene, A. (2007). *Cleaner Production Technologies and Tools for Resource Efficient Production.* The Baltic University Press.

[70] Zainon Noor, Z. (Ed.). (2012). *Introduction to Cleaner Production.* http://cp.doe.gov.my/givc/wp-content/uploads/2012/05/Introduction-to-Clean-Production.pdf (accessed 07.02.2018).

[71] Hilson, G. (2003). Defining "cleaner production" and "pollution prevention" in the mining context. *Minerals Engineering* 16 (4): 305-321.

[72] Lennart Nilsson, Per Olof Persson Lars Rydén, Siarhei Darozhka and Audrone Zaliauskiene, 2007: Cleaner Production Technologies and Tools for Resource Efficient Production. The Baltic University Press © 2007.

[73] UNIDO/UNEP (2010). Enterprise-Level Indicators for Resource Productivity and Pollution Intensity A Primer for Small and Medium-Sized Enterprises.

[74] Reproducing Calculations for the Analytical Hierarchy Process. http://professorforman.com/Reproducing%20AHP%20calculations.pdf (accessed 12.02.2018).

[75] Roy, B. (1981). The optimisation problem formulation: Criticism and overstepping, *Journal of the Operational Research Society* 32: 427-236.

[76] Kochov, A. and Mladenovska, D. (2015). "Identification of technical indicators for creating natural gas supply policies – Balkan case," invited lecture for the European Commission JRC & the Energy Community Secretariat Joint Workshop on Energy Scenarios for South Eastern Europe, 15 Dec, 2015, Vienna.

[77] Ishizaka, A. and Nemery, P. (2013). *Multi-Criteria Decision Analysis: Methods and Software*, John Wiley & Sons, UK.

[78] Nemery, P., Ishizaka, A., Camargo, M. and Morel, L. (2012). Enriching descriptive Information in ranking and sorting problems with visualizations techniques, *Journal of Modelling in Management* 7 (2): 130-147.

[79] ten Brink, B. (2006). A Long-Term Biodiversity, Ecosystem and Awareness Research Network, Indicators as communication tools: an evolution towards composite indicators.

[80] Saaty, T. L. (1977). A scaling method for priorities in hierarchical structure. *Journal of Mathematical Psychology* 15 (3): 234-281.

[81] Graphical and technical options in Expert Choice for group decision making. DTU Transport Compendium Series part 3. Department of Transport, Technical University of Denmark. First Edition, 2014.

References

[82] Hunjak, T. and Begičević, N. (2006). Prioritisation of e-learning forms based on pair-wise comparisons. *Journal of Information and Organizational Sciences* 30 (1).

[83] Saaty, T. L. and Vargas, L. G. (1991). *The Logic of Priorities, The Analytic Hierarchy Process Series.* Vol. III, RWS Publications, USA.

[84] Adamović, P., Dunović, Č. and Nahod, M.-M. (2007). Expert choice model for choosing appropriate trenchless method for pipe laying; TECHSTA 2007, 5th *International Conference*, pp. 19-31, September 2007, Czech Technical University of Prague.

[85] Graphical and technical options in Expert Choice for group decision making: DTU Transport Compendium Series 3; Department of Transport, Technical University of Denmark, 2014.

[86] Tsagdis, A. (2008). The use of the Analytical Hierarchy Process as a Source Selection Methodology and its Potential Application within the Hellenic Air Force.

[87] Cochrane, J. L. and Zeleny, M. (1973). Compromise programming in multiple criteria decision making, University of South Carolina Press, Columbia.

[88] Kamakura, W. A. (1991). Estimating flexible distributions of ideal-points with external analysis of preferences. *Psychometrika* 56 (3): 419–431.

[89] Danielsson, P.-E. (1980). Euclidean distance mapping. *Computer Graphics and Image Processing* 14 (3): 227-248.

[90] Aggarwal, C. C., Hinneburg, A. and Keim, D. A. (####). On the Surprising Behaviour of Distant Metrics in High Dimensional Space. https://pdfs.semanticscholar.org/f9af/c4590ac7288e722bc154cbfd73be1f575b58.pdf (accessed 20.02.2018).

[91] Alimov, A. R. (2004). Characterisations of Chebyshev sets in Co. *Journal of Approximation Theory* 129 (2): 217-229.

[92] Hwang, C. L. and Masud, A. S. M. (1979). Multiple objective decision making: Methods and Applications: A state of the art survey, *Lecture Notes in Economics and Mathematical Systems*, Vol. 164, Spring-Verglag, Berlin.

[93] Gass, S. I. and Roy, P. G. (2003). The compromise hypersphere for multiobjective linear programming. *European Journal of Operational Research* 144: 459-479.

[94] De, P. K. and Yadav, B. (2011). An algorithm for obtaining optimal compromise solution of a multi objective fuzzy linear programming problem, International *Journal of Computer Applications* 17: 20-24.

[95] Sioshansi, F. (2010). *Generating electricity in a carbon constrained world.* Elsevier.

[96] The impact of the financial and economic crisis on Global energy Investment. OECD/IEA, Paris: 2009.

[97] International Atomic Energy Agency. Integrated Energy planning for Sustainable Development. https://www.iaea.org/OurWork/ST/NE/Pess/assets/IEPSD%20Brochure%20WEB.pdf (accessed 21.02.2018).

About the Authors

Fisnik Osmani, PhD, born 1988, BSc (2010) and MSc (2013) in Mechanical Engineering Faculty – University of Prishtina. PhD (2018) from the Sts. Cyril and Methodius University - Faculty of Mechanical Engineering in Skopje – North Macedonia. PhD thesis title: "Defining indicators for decision and policy making for introducing technologies of cleaner production for contributing the sustainable development".

Dr. Osmani is an Assistant Professor at the University Isa Boletini – Mitrovicë, Faculty of Mechanical Engineering and Computing/Department of Production Engineering and Engineering Economy in the bachelor and master program. He has participated in several international conferences, and is also the author and co-author of many scientific and professional articles, projects and textbooks.

Email: fisnik.osmani@umib.net

Atanas Kocov, PhD, born 1966, BSc (1990) and MSc (1994) in Mechanical Engineering, Doctoral study program University of Washington 1995-1996; PhD (2002) from the Sts. Cyril and Methodius University in Skopje; PhD thesis title: "Theoretical and experimental investigation of composite materials and their implementation in metal forming tool design." Dr. Kocov is a former US Fulbright Scholar - postdoctoral study program at the University of Washington (Seattle, USA). Dr. Kocov was Dean of the Faculty of Mechanical Engineering in Skopje, where he has been working since 1991. Dr. Kocov was a national coordinator and Director of the UNIDO center and National Cleaner Production Center (NCPC), which was established in 2006 and still conducts on-going activities in Macedonia. The Center is implementing CP, RECP, low carbon technologies, circular economy and chemical leasing methodology in SMEs.

Email: atanas.kochov@mf.edu.mk

Index

A

Analytical Hierarchy Process – Repeated (AHP), v, vi, vii, 15, 16, 18, 19, 21, 40, 41, 42, 44, 48, 49, 83, 87, 90
availability of energy (AE), 61, 88

B

Balkans, 3, 29, 87, 89, 90, 92

C

clean energy, 4, 5, 7, 23, 76, 88
cleaner production (CP), v, vii, 1, 2, 3, 22, 28, 29, 31, 33, 35, 44, 49, 50, 83, 85, 87, 88, 89, 90, 94, 95, 97, 98
corruption control (CC), 10, 24
costs (C), 22, 23, 24, 32, 74, 84, 88, 89, 92, 93, 94, 95, 96

D

developed countries, 2
developing countries, 31
direct foreign investment, 24
drinking water, 35

E

Earth Summit, 4
economic growth (EG), 1, 51, 85, 88
electricity, 22, 24, 35, 37, 74, 75, 78, 83, 84, 85, 94, 96
emission, 6, 58, 76, 78, 84, 87
energy, 1, 2, 3, 4, 5, 7, 8, 9, 14, 16, 17, 22, 23, 24, 28, 29, 31, 32, 34, 35, 37, 51, 53, 75, 77, 78, 80, 81, 83, 84, 85, 87, 88, 89, 92, 94, 96
energy conservation, 31

energy consumption, 2, 29, 34
energy efficiency (EE), 28, 31, 34, 45, 51, 85, 88, 89
energy security, 87
energy supply, 1, 85
environment, 1, 2, 3, 4, 5, 6, 7, 16, 24, 31, 32, 33, 34, 38, 83, 84, 88, 91
environmental degradation, 5, 92
environmental impact, 28, 83
environmental issues, 33
environmental policy, 33
environmental protection, 5, 29, 83

F

foreign direct investment, 7, 51

G

good governance (GG), 53, 57, 86, 88

H

Human Development Index (HDI), 10, 24, 86

I

Internal Rate of Return (IRR), 9, 23, 53
International Atomic Energy Agency, 96
International Energy Agency, 92

K

Kosovo, vii, 1, 2, 3, 5, 7, 8, 22, 28, 37, 45, 51, 83, 84, 85, 87, 88, 89, 90, 92
Kyoto Protocol, 93

L

landscape changes (LC), 53, 60, 84

M

Multi Criteria Decision Analysis (MCDA), 41
Multi Criteria Decision Making (MCDM), 93

N

National Renewable Energy Laboratory, 92
natural gas, 7, 94, 95
natural resources, 3, 4, 32

O

OECD, 91, 92, 96

P

political stability and absence of violence (PV), 10, 24, 86

Q

Quality of Life (QL), 9, 57, 86, 88

R

random index (RI), 20, 21
raw materials, 33, 35
resource efficiency (RE), 1, 2, 3, 28, 44, 50, 51, 83, 84, 88, 89, 90

resource efficiency and cleaner production (RECP), v, 1, 2, 31, 32, 94, 95, 98
rule of law (RL), 10, 24, 86

S

safety and health (SH), 51, 56, 85, 87, 88
securing clean energy (SCE), 51, 53, 60, 85, 88
security of supply (SS), 61, 88, 89
Small and Medium Enterprises (SMEs), 29, 37, 84, 85, 87, 88, 98

T

thermal energy, 8, 85
thermal power plant (TPP), 5, 28

U

United Nations (UN), 2, 3, 4, 5, 7, 31, 88, 91, 92
United Nations Environment Programme (UNEP), 31, 32, 94, 95
United Nations Industrial Development Organization (UNIDO), v, 31, 34, 94, 95, 98

V

voice and accountability (VA), 10, 24, 86

W

waste treatment (WT), 53, 58
Western Balkans (WB), 3, 29, 87, 90, 92